INVESTMENT
AND AGRICULTURAL DEVELOPMENT
IN DEVELOPING COUNTRIES

INVESTMENT
AND AGRICULTURAL DEVELOPMENT IN DEVELOPING COUNTRIES

The Case of Vietnam

CUONG TAT DO

August 2015

Copyright © 2015 by Cuong Tat Do.

Library of Congress Control Number: 2015919567
ISBN: Hardcover 978-1-5144-4272-2
Softcover 978-1-5144-4273-9
eBook 978-1-5144-4274-6

All rights reserved. No part of this book may be reproduced or transmitted in any form or by any means, electronic or mechanical, including photocopying, recording, or by any information storage and retrieval system, without permission in writing from the copyright owner.

Any people depicted in stock imagery provided by Thinkstock are models, and such images are being used for illustrative purposes only.
Certain stock imagery © Thinkstock.

Print information available on the last page.

Rev. date: 12/02/2015

To order additional copies of this book, contact:
Xlibris
1-800-455-039
www.Xlibris.com.au
Orders@Xlibris.com.au
710374

CONTENTS

ABSTRACT xi
DEDICATION xv
ACKNOWLEDGEMENTS xvii

CHAPTER 1 INTRODUCTION 1

1.1. The types of capital 2
1.2. Motivation and scope of research 9
1.3. Research overview 20
1.4. The contributions of the study 35
1.5. The structure of the thesis 40

CHAPTER 2 HISTORICAL CONTEXT AND POLICY DEBATES 43

2.1. Introduction 43
2.2. Stages of agricultural development worldwide 47
2.3. Actions for agricultural development in Vietnam 53
2.4. Policy debates and new directions for agricultural development in the future 75

CHAPTER 3 DATA DESCRIPTIONS 81

3.1. Vietnam Household Living Standard Survey 82
3.2. Vietnam Enterprise Survey 93
3.3. Provincial Competitive Index Survey 95

CHAPTER 4 HUMAN CAPITAL, SOCIAL CAPITAL, AND AGRICULTURAL DEVELOPMENT: A LITERATURE REVIEW 97

4.1. Introduction 97
4.2. Human capital 98
4.3. Social capital 116
4.4. Agricultural development 135
4.5. Conclusion 147

CHAPTER 5 THE DETERMINANTS OF HEALTH INVESTMENT TO AGRICULTURAL DEVELOPMENT 149

5.1. Introduction 149
5.2. Literature review 151
5.3. A basic conceptual framework 160
5.4. Methodology 169
5.5. Data and variables statistics 172
5.6. Estimations and results 178
5.7. Policy implications 188
5.8. Conclusions 192

CHAPTER 6 EDUCATIONAL INVESTMENT AND AGRICULTURAL DEVELOPMENT IN VIETNAM 195

6.1. Introduction 195
6.2. A brief review of education in Vietnam 198
6.3. Theoretical framework 204
6.4. Methodology 210
6.5. Data summary 212
6.6. Empirical results 217
6.7. Policy implications 230
6.8. Conclusion and further research 231

CHAPTER 7   SOCIAL CAPITAL AND AGRICULTURAL DEVELOPMENT IN VIETNAM 233

7.1. Introduction 233
7.2. Background information on social capital related to agricultural development in Vietnam... 235
7.3. Theoretical framework 244
7.4. Methodology 252
7.5. Empirical results.. 263
7.6. Policy implications 285
7.7. Conclusion.. 287

CHAPTER 8   PUBLIC POLICY FRAMEWORK FOR FURTHER AGRICULTURAL DEVELOPMENT 289

8.1. Introduction 289
8.2. Public policy implications on human capital development..... 292
8.3. Public policy implications on social capital development 301
8.4. Conclusion..311

CHAPTER 9   CONCLUSION..314

REFERENCES 333

TABLES AND FIGURES

Table 1.1: The expenditure poverty rate using the
World Bank poverty line 14

Table 3.1. Average expense on education and training per person in the
past 12 months by expense item, urban rural, region......... 83

Table 3.2. Economically active population in working age by urban,
rural, income quintile in 2002, 2004, and 2006 87

Table 3.3. Monthly income per capita by source of income and
its structure by urban, rural in 2002, 2004, and 2006 89

Table 3.4. Average health expenditure per person having treatment
in the past 12 months by type of treatment,
urban and rural region 91

Table 3.5. General information of agricultural enterprises
from 2000 to 2008 94

Table 4.1. Rural poverty rate worldwide 1993–2002............................ 137

Table 5.1. Summarises the different characteristics across the three
household surveys 173

Table 5.2. Income components 175

Table 6.1: Percentages of population aged 15 and above by
highest certificate and income quintile in 2006 203

Table 6.2. Statistical summary 215

Table 6.3. Statistical description of income and education
among regions 216

Table 6.4. Determinants of agricultural households' income
(cross-sectional sample) 219

Table 6.5. Determinants of agricultural households'
income (panel sample) 221

Table 6.6. Contribution of investment in education to
    agricultural households' income 224
Table 6.7. Rates of return on educational investment
    across regions in Vietnam 227
Table 7.1. The effect of social capital on agricultural households'
    income in 2002 263
Table 7.2. The effect of social capital on agricultural households'
    income in 2004 267
Table 7.3. The effect of social capital on farmers' incomes in 2006 270
Table 7.4. The summary of effect of social capital on agricultural
    household incomes in 2002, 2004, and 2006 274
Table 7.5. Result of estimating the baseline model 275
Table 7.6. Estimated the effect of weather on agricultural firms 276
Table 7.7. The effect of social capital on productivity of agricultural
    firms at provincial level from 2005 to 2008 277
Table 7.8. Estimated results of the time-varying decay
    inefficiency model 281
Table 7.9. Effects of social capital on technical inefficiency 283
Table 7.10. The summary of the effect of social capital on technical
    inefficiency of agricultural firms from 2005 to 2008 at
    provincial level 284
Table 5.2(a) Summarize statistic of variables in VHLSS 2002 369
Table 5.2(b) Summarize statistic of variables in VHLSS 2004 370
Table 5.3. Estimation result by using VHLSS 2002 373
Table 5.4. Estimation result by using VHLSS 2004 374
Table 5.5 Estimate the speed of knowledge diffusion
    by using VHLSS 2002, 2004 375
Table 5.6. The effect of investing in health on income of agricultural
    households in 2002, by regions 377
Table 5.7. The effect of investing in health on income of agricultural
    households in 2004, by regions 378
Table 5.8. Estimates of the effect of health on economic growth a 379

Table 7.11. Statistical summary of variables 382

Figure 2.1: The density of rural populations of Vietnam
from 1930 to 2010 57
Figure 2.2: The density of rural labour of Vietnam from 1990 to 2010 57
Figure 2.3: Major agricultural and rural financial providers in Vietnam 64
Figure 2.4: Major institutions of the rural financial market in Vietnam 67
Figure 5.1: The profile of output across line of different vintage,
at date t 164
Figure 5.2: The timeline of health status of the worker 166
Figure 6.1: The education system in Vietnam 200
Figure 6.2: Trends in School Enrolment, 1987–2010 201
Figure 5.2. Frequency distributions for log agricultural households
and food expenditure income in 2002 371
Figure 5.2. Frequency distributions for log agricultural households
and food expenditure income in 2004 372

ABSTRACT

The standard policy prescription to enhance the productivity of agriculture in a developing country like Vietnam is first to encourage the investment of farmers in their human and social capital and then to change the governmental institutions to facilitate the farmers' investment. This thesis, therefore, analyses the investment of farmers in their health, education, and social relationships in the context of Vietnam's recent agrarian transition. Using the tools of regression analysis, the author has tried to measure the rate of return of investment in health, education, and social relationship of farmers on their income. Additionally, to measure the effect of local government policy on the performance of agricultural firms at the provincial level, the thesis applies current techniques to estimate the relationship between output of agricultural firms and performance of local government. As in other low-developed countries, the rate of return of investment in education is quite extreme and the rate of investment in health is small. The effect of investment in social capital of farmers on their income is quite complex due to the complication of the term social capital. Moreover, the effect of quality of local institutions on performance of agricultural firms at the provincial level is not quite statistically significant. The relationship is similar with other research findings worldwide when researchers want to measure the effect of quality of institution on the performance of firms.

The way in which investment activities was assigned to agricultural households in the first stage of transition—in particular, the rights of farmers to invest privately in their land—was clearly crucial both for improving living standards of farmers and the performance of economic development. However, the heavy reliance on physical investment has raised a concern about horizontal investment—which it leads to an unstable increase of agricultural productivity. Vertical investment with focal point on technology, skills, and ability of farmers, and quality health farmers is needed for a modern development of agriculture in developing countries. Evidently, in Vietnam, horizontal investment did help to develop agriculture in the past, while now this way of investment is binding the room of further development of agriculture. This thesis has first tried to see if such concerns are borne out by the evidence on how investment in health, education, and social capital affect farmers' income under the new rural development program introduced by Vietnam since 2000. This was arguably the most crucial task in the country's transformation to a market-based agriculture after abandoning the inefficient collective system. Individual households had to be assigned the rights to invest on their own land. The author has used a model of household income to analyse the effect of investment in health and education on their rights of investment on their own land, expressed by their income. The results are quite similar with the picture that many commentators have painted so far.

The contributions of the thesis could be accounted mainly two folds. *Firstly*, this is the first time that the relationship between investment in human and social capital and income of agricultural households in Vietnam has been analysed comprehensively. There are lots of researches on the relationship between education and income and analysis the effect of health insurance program on Vietnam households or poor households, but there is hardly to find a research to analyse the effect of investment in education, health on households' income, especially agricultural households. *Secondly*, analysing the effect of investment in social capital from micro to macro level could be seen as my thesis's second contribution. Investment in social capital at micro level includes time allocation for bonding, bridging relationships among farmers and their money to organize party where farmers enhance their relationship; at macro level includes the changing of local government policy in privatization incentive, human resource development, or state-owned enterprises priority policy. The empirical results for micro level shows that investment in social capital has contributed partly positively to agricultural households' income, while at macro level the effect of investment in social capital is quite complex. The complex outcome of empirical results is due to the complicated definition of social capital. Social capital cannot define itself in one definition. Social capital is a multi-faceted definition. Therefore, it is hardly to capture its effects to agricultural households' income.

DEDICATION

This book is dedicated to my grandfather in his wonderland, Khuy. My memories of your invaluable love and affection, tolerance, confidence and belief in me are truly appreciated. I love and miss you with all my heart.

ACKNOWLEDGEMENTS

This dissertation is the outcome of my work at the Faculty of Business, Government, and Law, University of Canberra, under the sponsor of Vietnamese Government, Project 322, for the period of 2008–2012. Many people have been of great help in completing this dissertation. I would like to thank all of them, both in person and in these acknowledgements. Unfortunately, it is impossible to mention all on this page.

I am grateful to Assistant Professor Greg Mahony for giving me invaluable advice on so many issues. Many special thanks are devoted to Associate Professor Cameron Gordon, who has been so helpful that I could not have finished my dissertation without his insight comments and advices. Moreover, I would like to thank to Dr Ben Freyens for his constructive advices and for reading my manuscript, although he is lately joining to my supervisory panel.

I like to extend my gratitude to Professor Phil Lewis, Dr Craig Applegate, and other members of Faculty of Business, Government, and Law for their invaluable comment on chapter 5 of the dissertation. Many special thanks are devoted to Professor Deborah Blackman for her feedbacks on chapter 6 and chapter 7 when I presented the two chapters at Hot Knowledge Conference, University of Canberra.

I also wish to say thanks to administrative members of Faculty of Business, Government, and Law, University of Canberra, who provide a very good support to my research. The supports from the faculty and the university in regard to the supplying a computer with internet connection, accessing to journal database, office place are appreciated. Without their help, I cannot complete my dissertation on time.

Last, but foremost, I would also like to extend my deepest gratitude to my grandfather in his heaven and my parents for teaching me how behave in decent manner and for encouraging me to pursue my education as far as possible. Many special thanks are dedicated to my wife who supports me a lot during the time of writing the dissertation. Without her support, I do think that I cannot complete timely the dissertation. This entire dissertation is dedicated to them.

Chapter 1
INTRODUCTION

The transformation from socialist command economy to a market-based economy brings both opportunities and risks for Vietnamese citizens. Like other poor economies, the initial transformation of Vietnam's economy began in rural areas where most citizens were poor. The process of economic development will typically move many rural and agricultural households out of farming and into remunerative (urban and rural) non-farming activities. Reformation shifts rural economies from controlled farming institutions under socialist agriculture to a more flexible, market-based economy where incentives are strong and where agriculture can then play a significant role in the process of economic growth. However, this transformation presents a major challenge for policymakers in that farmers do not have the necessary skills required for working in the industrial sector. As a result, income differences between farm households and non-farm households are widened and the living standards of farmers become viewed as socially unacceptable.

In order to solve the problems related to transformation, the Vietnamese government has implemented nationally targeted programmes aimed at improving farmers' skills and knowledge. Human and social capital (in the form of social infrastructure) has been addressed not only by the Vietnamese government but also by the international community. There are many projects and programmes in rural areas whose main aim is to raise the abilities of farmers and have their voices heard in society. However, projects and programmes sponsored by the international community are only pilot activities and do not have any effect in terms of the country as a whole. The policy problem in this instance is that investment in human and social capital needs to be comprehensively addressed by the government.

This research studies how investments in human and social capital at the household level affect the incomes of farmers. The empirical results of the analysis will be a sufficient input for the government in implementing nationally targeted programmes. This chapter first reviews the types of capital needed, then provides an overview of the scope of the research, proposes the research questions and hypotheses, thesis chapters, and the limitations of the research.

1.1. The types of capital

In the current economic system, there are essentially three types of capital that investors can select: physical, human, and social capital. Investors can choose to invest in one or all types of capital, depending on their goals and financial resources. Normally, firms will choose to invest in physical and human capital, while individuals and governments choose to invest in all types of capital. However, currently, we are experiencing a change in the way economic agents invest in capital. Alongside governments, firms now also invest in three types of capital. Individuals generally know more about the three types of capital and are therefore willing to invest in all of them if they have adequate financial resources.

Physical and human capital has long been part of economic theory and empirical economics, while social capital has only recently been employed by economists. Physical capital includes all tangible capital, such as machineries, tractors, and oil wells. Human capital refers to all intangible individual abilities and knowledge. It is usually approximated by years of schooling or health status. Social capital refers to the relationship between individuals at the micro level and institutions at the macro level. Social capital can be understood as the shared knowledge, understanding, norms, rules, and expectations that drive the actions of groups or individuals to achieve their common interests. A comprehensive literature review on human and social capital will be presented in chapter 4.

The effects of physical capital and human capital on economic development have been widely explored not only at the micro level, but also at the macro level. Additionally, the differences between countries regarding physical and human capital accumulation can be seen as a major reason for cross-country income differences. The effect of social capital on economic development, on the other hand, has just recently been analysed by economists. Developing countries are focusing on physical and human capital as two major workhorses for escaping low development status, while social capital has received less attention. The reason is that the effects of physical and human capital on income per capita and economic growth rate are quite clear and significant, whereas the effect of social capital is not usually significant. However, not all economists refute the effect of social capital on the development process. Therefore, two questions arise: (1) What types of capital should be invested in by developing countries? (2) What resources should be invested in for each type of capital?

It should be emphasised that the present study is concerned with the micro level of the above questions. That is, we ask if it is formally possible to construct a list of investing items from the set of known individual potential investment activities, the procedure in question being required for certain natural conditions. Suppose there is an individual endowed with three types of capital and the individual must choose one or more investment items from these three types of capital (e.g., physical capital, human capital and/or social capital). It is expected that the individual can invest repeatedly, but sometimes not all of the three types of capital will be available. Similarly, in the usual investment project analysis and cost-benefit situations, rational investment behaviour regarding the three types of capital would mean that the individual orders the three types of capital according to his collective preference and cost-benefit calculation. Additionally, based on their expectations regarding the future value of the present value, investors will select any given case that type of capital among those available with highest benefit on the list. Naturally, one type of capital will be invested in rather than another if the type is expected to yield a higher return and is suitable with the current status of wealth. Let *A*, *B*, and *C* be the three investment projects, where *A*, *B*, and *C* are the list of all available types of capital (e.g., may have one or more types of capital), and where *1*, *2*, and *3* are three individuals. Suppose individual *1* prefers *A* to *B* and *B* to *C* (and then prefers *A* to *C*), individual *2* prefers *B* to *C* and *C* to *A* (and then *B* to *A*), and individual *3* prefers *C* to *A* and *A* to *B* (and then

C to *B*). In this instance, the majority trend of investment activity prefers *A* to *B* and *B* to *C*. If the community follows rational behaviour, we would have to say that *A* is preferred to *C* and all individuals will invest their money in the same list of investment projects. However, the major trend in investment does not function in this manner.

Additionally, because all individuals have to consider their budget constraints, they need to carefully calculate their scope of investment for each type of capital. In order to estimate the optimal list of investment activities, individuals have to compute their budget based on their available financial sources. The availability of own resources, loans from banks, borrowing from relatives or other persons or agents, and earnings from sales should be identified.

The structure of investment is different from person to person. An optimal investment structure will guarantee that individuals have enough resources to finance their investment. Highly educated persons will have many estimating tools available to them, or can learn them at little cost, while less-educated persons may not have such tools available or are unable to gain that knowledge for various reasons. Therefore, less-educated persons and poor people need instruction from more knowledgeable investors or government to facilitate their investment decisions.

Physical capital and human capital can be invested in by individuals and government. Investment in large infrastructure projects is important for improving individuals' income and central government should fund this type of investment, while smaller infrastructures such as a cross-village road, can be affected by villagers under the sponsorship of government.

Investment in education can be effected by either individuals or government. Gaining knowledge at school is the responsibility of individuals, while providing the necessary education facilities for doing so is the responsibility of the government. Individuals spend their time at school with the expectation that in the future they will be able to earn more money with their qualifications. Thus, the government body in charge of education should implement a realistic framework for building up curriculums at school. Teachers can make informed choices about textbooks only when there is enough choice for them within the textbook market. If the government does not support diversity within the textbook market, learners will experience fewer benefits from learning.

Investment in individuals' health can be done simply and cheaply at the individual level through exercise. For more complex investment in health, such as purchasing medication, medical tools, or hospital stays, individuals need to allocate money for using these facilities, while it is the responsibility of government to invest in building hospitals and creating pharmaceutical companies. Moreover, the government needs to maintain a competitive market for medicines and make sure that all individuals have access to this market. Improving health is quite different from gaining more knowledge, but there is nonetheless a link between them. People with better health will gain more knowledge and therefore have higher productivity.

'Social capital' is a complex term. Economists agree that this notion is multifaceted and that we cannot proxy social capital as one economic variable. Social capital can be seen as the relationship between members of a group or network. Alternatively, it might also be considered to represent institutions.

Social capital can be seen as trust; the complexity of the concept makes measuring social capital a challenge not only for economists but also for sociologists. There is much research in sociology and health about social capital; in economics, however, research in this area is lacking. Economic researchers only concentrate on institutions and try to explain their effect on economic growth or on the performance of economies. Indeed, the modelling of cooperative behaviour in the theory of industrial organisation can be seen as an example of social capital at the institutional level.

The working mechanism of social capital in society is extremely complex. For example, in a simple case, if we define a group as X, all members of this group are indexed by X_i ($i = 1...n$) and each member has an individual utility function called U_i. U_i is a function of gains from group interactions. The gains may be knowledge or opportunities for finding a better job or simply the benefits from expressing oneself in front of others. This means that each member of group X will gain some benefit when they join the group and the gained utility is different from person to person. However, members of group X should have some aspects in common, because their group is based to some degree on the commonality of their demand and utility. Their demand and utility should somehow overlap or interfere. Thus, utility overlapping builds up a spectrum, so that all members can benefit from the group. This effect is similar to Pareto efficiency. The benefit from group X is non-excludable. No one in the group can limit the chance of others benefitting from group X.

A more complicated situation occurs when a member of group X joins more than one group. Other groups may be Y, Z, or B and each group has its own members. Members of group X will interact with members of group Y, Z, or B and benefit from these groups. X_i will carry new knowledge or benefit from group Y, Z, or B to group X in order to share value to members of group X. Members of group Y, Z, or B can then do exactly what members of group X do; information thereby flows from one group to other groups via their members. The interaction between members of these groups therefore increases the ability of all members, thereby increasing their productivity through the amassed knowledge of all members.

Two questions arise here: What should government do to support investment in agriculture? How much should government spend for each type of capital? In terms of the first question, government should invest in three types of capital, because physical, human, and social capitals have positively affected agricultural production and agricultural households' income. Additionally, government should implement nationally targeted programmes (hereafter NTPs) that address agriculture. Through NTPs, the government can transfer funding based on the needs of farmers and let farmers themselves decide what should be invested in. The government should act as a supporter in the agricultural development process rather than a direct agent. Additionally, government activities should also entail an element of social capital. For instance, government can increase the productivity of current physical and human capital by introducing incentive policies such as reduced tuition fees or by providing cheaper healthcare services for farmers.

Regarding the second question, it is very risky to estimate exactly the fraction of investment in each type of capital, because it depends on the resolve of government for each stage of development. At the initial stage of development, government should invest more in physical capital. Next, it should aim to use its physical capital efficiently, focus on human capital needs, and then invest in social capital as a major stabilisation factor. It is, however, not necessary to divide the process into the three above stages. Physical, human and social capital can be invested in at each stage; alternatively, only one or two types of capital can be invested in, depending on the availability of financial resources. We cannot invest more in social capital in a specific area if the area already has a good social capital endowment. In urban areas, human capital will be invested in more than in rural areas, because of better financial resources. However, if the focus is on culture, rural areas could be worthy of higher investment, because these areas maintain and support traditional cultural values.

In short, each type of capital has its own characteristics and their working mechanisms are different. It is said that the process of investment in physical and human capital are quite clear and easier to conduct than investment in social capital. Physical and human capital can deteriorate moderately without maintenance, while social capital can deteriorate completely if we do not properly oversee it. The major part of physical capital is the physical object and the main element of human capital is individual ability, while the key component of social capital is the connection between individuals and groups. Physical objects can be damaged by weather or natural forces; human ability can be damaged by ill health or natural forces. Social capital, however, can be damaged only by human behaviour. Therefore, physical, human, and social capital can be accumulated and improved on only via human activities.

1.2. Motivation and scope of research

1.2.1. Motivation

In 1986, Vietnam launched a reformation strategy—a homegrown, political, and economic renewal campaign—called *Doi moi*, marking a transformation from a centrally planned economy to a socialist-oriented market economy. At the time, Vietnam was one of the poorest countries in the world and faced many challenges such as hyperinflation, famine, drastic cuts in Soviet aid and a trade embargo by the West.[1] For most Vietnamese, life was unpleasant and the future seemed bleak. When measured against this backdrop, the performance of Vietnam's economy over the last two decades has been remarkable. From 1990 to 2010, Vietnam's economy grew 7.3 percent annually and income per capita was almost five times as much. The rapid expansion of the economy has been accompanied by a high level growth of international trade, large-scale foreign direct investment, a significant reduction of poverty, and almost universal access to primary education, healthcare, and infrastructure such as paved roads, electricity, piped water, and housing. Consequently, Vietnam's transition from a centrally planned economy to a market economy and from an extremely poor country to a lower-middle-income country within two decades is now a case study in economic development textbooks.

During the early stages of *Doi moi*, one important feature was the relative importance of the rural sector and the dominant role of household units in Vietnam's agricultural production. Arkadie and Mallon (2003), Lin (2009), and others have argued that Vietnam, like China, was largely an agrarian country at the beginning of its transformation, so its production was broadly consistent with its comparative advantage. Therefore, when Vietnam opened its economy, the agricultural sector responded strongly

---

[1] In 1990, income per capita in Vietnam was only US$ 98 (in current US dollar). Vietnam was indeed the poorest country. Another two poorest countries were Somalia and Sierra Leon with income per capita were US$ 139 and US$ 163, respectively. Vietnam was among 20 poorest countries in terms of gross domestic product per capita adjusted for purchasing power parity.

to the changes that promoted the development of agriculture, offsetting all contractions in the industrial sector. For instance, Resolution 10 of the Party, in 1988, provided farmers with property rights (albeit limited), which was arguably a critical point in Vietnam's agricultural development. With limited property rights, farmers could invest in their physical, human, and social capital in order to obtain a better living standard. In fact, the limited property rights and the reformation of price and trade contributed significantly to sustaining agricultural growth, and generated the surplus necessary for diversifying into a non-agricultural sector, thereby strengthening the flexibility of the economy.[2]

In the past, agrarian collectivisation had been an important ingredient of the socialist strategy in Vietnam. This was particularly true in the North, where cooperatives were developed as productive units and service providers. The experience in the South was different from that of the North. There were two successive waves of collectivisation in the South, one in 1979–1980 and the other in the early 1980s, but collectives never played as important a role in the southern rural economy as they did in the North. As has been documented, the transformation in the agricultural sector was mainly started by the resistance of farmers in the South. Particularly, farmers in the South refused to grow rice beyond their families' requirements. Several senior policymakers witnessed the benefits of household farming and later formulated policies to encourage similar changes throughout the country (Dixon, 2003; Rama & Võ, 2009). They de-collectivised agriculture, established land-user rights, reduced

---

[2] The other factors are the timing of natural resources exploration such as oil and coal, and Vietnam's location in one of the most dynamic and fastest growth regions in the world.

the role of cooperatives, and liberalised agricultural prices, transforming the country in two years from being chronically food scarce to the third-largest exporter of rice.

The transformation of agriculture in Vietnam has four characteristics that are similar to common global trends. *Firstly*, agriculture employs far more people than other industries and sectors; around 60 percent or more of total labourers can be found employed in agriculture, while in developed countries, this number is less than 10 percent. *Secondly*, the rural economy in Vietnam is described as tradition bound. In the industrial sector, products can be manufactured only by applying modern technology and engineering, but in agriculture, ways of producing often use techniques developed hundreds or even thousands of years prior to the arrival of modern technology and science. Additionally, rural communities in Vietnam, in which traditional techniques are often employed, have customs and attitudes that support older ways of doing things, and therefore inhibit change. *Thirdly*, land and weather are two important factors affecting production. Land and weather differ from place to place, so that techniques suitable in one place cannot necessarily be applied successfully in another. *Finally*, agriculture is the only sector that produces food. Vietnamese people, and more generally humankind, can survive without coal and electricity but not without food. Unlike most manufactured products, for which there are substitutes, food is not substitutable. Food must either be produced within a country or imported.

Agriculture's role in economic development in Vietnam is central, because most Vietnamese make their living from the land. If leaders are seriously concerned about the welfare of their people, the only way for them to increase this welfare is by supporting, *firstly*, an increase in farmers' productivity in growing food and cash crops, and *secondly*, in the prices they receive for their crops. However, not all increases in farm output obviously benefit the majority of rural people. Development of mechanised, large-scale farms in place of small, peasant farms might actually render the majority of rural people worse off. Although raising farmers' productivity is a necessary condition, raising agricultural output is not by itself sufficient for achieving an increase in rural welfare.

The size of the agricultural sector is an important characteristic that offers an explanation for the development of other modern sectors. With a dominant portion of the population in agriculture, the rural sector is virtually the only source of increased labour for the urban sector; however, it may constrain the development of the urban sector if the quality of labour in the rural sector is not good enough. Rural workers can transfer to an urban economy only if they can meet the work requirements thereof. Unfortunately, the quality of rural workers usually cannot meet the needs of an urban economy. If this restriction is not removed, economic development will be severely crippled. The current situation of the relationship between agricultural and industrial sectors regarding the quality of labour in Vietnam fits this statement. This factor is therefore the first motivation for doing this research.

A reform strategy, for particular, was implemented more than two decades before the starting point of this thesis. Nonetheless, rural households currently still outnumber poor households in Vietnam. The summary of poverty rates using the World Bank poverty line is provided in the following table.

Table 1.1: The expenditure poverty rate using the World Bank poverty line

Unit: percent

	1998	2002	2004	2006	2008
1. The entire country	*37.4*	*28.9*	*19.5*	*16.0*	*14.5*
- Urban	9.5	6.6	3.6	3.9	3.3
- Rural	44.9	35.6	25.0	20.4	18.7
2. Regions					
- Red River Delta	30.7	21.5	11.8	8.9	8.0
- Northern midlands and mountainous area	64.5	47.9	38.3	32.3	31.6
- North central and central coastal area	42.5	35.7	25.9	22.3	18.4
- Central Highlands	52.4	51.8	33.1	28.6	24.1
- South East	7.6	8.2	3.6	3.8	2.3
- Mekong River Delta	36.9	23.4	15.9	10.3	12.3

Source: GSO (2010)

Above all, the expenditure poverty rate decreased significantly from 1998 to 2008. The northern midlands and mountainous regions have had the highest reduction; the rate declined by 32.9 percent during this period. The next region is Central Highlands with a reduction of 28.3 percent during the same period. The other regions also performed significantly in reducing the poverty rate. Additionally, the poverty rates in both rural and urban areas decreased significantly. The rate fell by 6.2 and 26.2 percent in urban and rural areas, respectively.

However, the majority of Vietnam's population lives in rural areas where the poverty rate remains high. This means that poor households in rural areas still remain more prevalent than poor households in Vietnam in general. Why does this remain the case after such a remarkable period of economic growth in Vietnam for more than two decades?

Practically, the agricultural sector might be seen as the most important source of capital for modern economic growth, especially for the early stages of development in a developing country like Vietnam. Capital comes from invested savings and savings from income. The question, however, is that how can rural people save their incomes? This is an important question for leading policymakers in Vietnam. A large number of rural people do not have any savings, because their incomes are just enough to support themselves and their families. Investment in health and vocational education can therefore be seen as an appropriate approach for assisting the rural population. In economic literature, health and education have been identified as two major inputs for productive workers.[3] Poor health and education is likely to affect the economic well-being of the rural population, while limited economic resources presumably affect health and education investments. Additionally, there are possible reactions between health, education, and resources.

To further support agricultural development in the coming decades, the Vietnamese government faces three options: (i) applying a mechanical package; (ii) employing a biological package; (iii) applying a combination of mechanical and biological packages. Mechanisation implies the application of large tractors and combines metal silos with mechanical loading devices and several other pieces of expensive equipment, while a biological package involves mostly the greater utilisation of chemicals and improved plant varieties to further develop the productivity of land.

Mechanisation of agriculture in labour-abundant developing countries is primarily a substitute for labour. In Vietnam, as in China or India, there

---

[3] This statement will be discussed thoroughly in chapter 4.

are periods when the demand for labour exceeds its supply. When two rice crops are produced each year, for instance, the first crop must be harvested, fields prepared, and the second crop transplanted, all within a matter of a few weeks. Transporting the harvest to market also needs a considerable amount of labour if goods must be delivered on carts hauled by animals or men, or as head loads carried by women. This is the case in developing countries. One driver with a truck can do in a day what might otherwise take dozens of men and women several days to accomplish.

In the case of Vietnam, hand tractors and rice transplanters are too expensive for most farmers to afford, despite the fact that the Vietnamese government removes all tariffs on imports. Farmers need more support from the government in order to be able to use this equipment in producing goods. In the absence of further support, farmers would be better off economically using labourers.

The main aim of the biological package, on the other hand, is to raise yields. There is nothing new about using improved plant varieties in combination with fertilisers and pesticides to raise yields of rice and/or corn. In the developing world, a rapid increase in the application of chemical fertilisers accompanied the increased use of high-yielding and other improved plant varieties. Unlike machinery, chemical fertilisers can be purchased in almost any quantity and even only small amounts can boost yields. Therefore, chemical fertilisers are the first priority of poor peasants. The limitations of using greater amounts of chemical fertilisers have thus far not been the conservatism of peasants or their poverty, but the availability of supplies and their prices.

A key component of the biological package is water. Improved plant varieties using more chemical fertilisers lead to significantly higher yields only if there is a sufficient and timely water supply. In Vietnam, rainfall provides all the water required and usually at the right time. In the event rainfall comes at the wrong time, farmers have water reservoirs. As a result, efforts to raise yields in Vietnam have usually focused on the extension of irrigation systems, so that crops are less dependent on the vagaries of the weather.

The extension of irrigation systems has been seen as primarily a financial and engineering problem. Vietnam has enough money, from aid and its own resources, to build dams to create reservoirs and canals to take water to the fields. However, the problem with running irrigation systems is that engineers can build the dams and the main canals, but they cannot involve farmers in building and maintaining the feeder canals to the fields. The problem sometimes becomes one of conflicting interests between farmers and the local politicians of a rural society. Therefore, irrigation extension in Vietnam is as much a social as an engineering, financial, and ecological obstacle.

More than anything else, the second motivation for this thesis is the notion that an individual or mixed package should be the focus of the Vietnamese government for encouraging further agricultural development in Vietnam in the future. The package selection will significantly affect policy ingredients. There is no doubt that the increased use of inputs from the biological package has made possible the steady expansion of agricultural output. In the future, the further development of improved plant varieties and the expansion of irrigation systems, together with increased chemical fertiliser production, will remain the major contributors to higher yields. In contrast, the main function of the mechanisation package continues to be freeing agricultural labour from the burden of producing food so that it can do other, ideally more productive tasks. If the government chooses the mechanisation package, it needs to spend more on physical objects and less on people. On the other hand, if the government chooses the biological package, it is required to spend more money on humans, because farmers need to train more in order to understand how fertilisers work for specific harvests. A considerable budget will have to be made available for environmental research regarding the use of chemical fertilisers. Furthermore, government spending on general education and healthcare will need to be taken into account, as in the case of Vietnam, the biological package relies on people. If the government chooses to combine these packages, an optimal portion of spending on each package should be carefully addressed.

1.2.2. Scope of research

In the present study, the subject of research is the relationship between investment and agricultural development. The relationship is analysed in order to establish suggestions for public policy.

To analyse this relationship, a sample of agricultural households throughout Vietnam was selected from three waves of the Vietnam Household Living Standard Survey (hereafter VHLSS). This survey was conducted by the Vietnam General Statistics Office under the technical support of the World Bank Vietnam and United Nations Development Program. The sample represents all agricultural households in Vietnam. Therefore, the conclusion of the research can be seen as a reliable source for implementing public policy in the field of agricultural development. The VHLSS followed two waves of the Vietnam Living Standard Survey (hereafter VLSS) in 1993 and 1998, which were conducted by the World Bank Vietnam. Unfortunately, there is no link between the VHLSS and the VLSS, as the goals of conducting both surveys were different. Thus, the research cannot accurately compare the two periods of 1993–1998 and 2002–2006. The research selected the VHLSS as the main dataset for analysing because it was available to the researcher. Additionally, the VHLSS questionnaire was developed mainly based on the VLSS questionnaire. This development was necessary because the VLSS questionnaire was simply an application of a questionnaire that hailed from other countries. The reason for using the VHLSS in the present study is that it is closer to the real situation in Vietnam than the VLSS.

As previously stated, there are three types of capital: physical, human, and social. The effect of physical capital on agricultural households' income and/or well-being has been comprehensively analysed in the case of Vietnam. However, research on the effect of human and social capitals on the incomes of agricultural household has not been as prevalently investigated by Vietnamese and international researchers. Kompas et al. (2012) provide a comprehensive analysis of the effects of land and market reform on rice output and incomes. The link between physical capital and output of rice production is illustrated using measures of total factor productivity, net incomes, and net returns in rice production from 1985 to 2006. The researchers found strong evidence that, in the past, land and market reform contributed considerably to gains in major rice growing areas. However, recent evidence obtained in this context indicates that productivity is slowing down. This situation can be seen as a frontier for exploiting physical capital; then, if the Vietnamese government wishes to further develop agriculture, it is strongly recommended that the government should focus on investment in human and social capital. Hence, for the present study to contribute to the current economic literature, the relationship between human and social capital and farmers' incomes will be the focal point. Physical capital has to date received only minimal research attention. Human capital, which is proxied by health and educational capital and social capital, which is proxied by inside-family and outside-family relationships, will be thoroughly analysed using the VHLSS datasets from 2002, 2004, and 2006.

Furthermore, this research addresses the correlation between human and social capital and agricultural development at the household level. The relationship is expressed by the link between farmers' incomes and their human and social capital. Particularly, the link between human health, education, and social capital and income has been estimated by using a production function approach at the household level. Each link will be described clearly in a specific chapter of the research.

1.3. Research overview

1.3.2. A summary of the literature review

This section will present a concise examination of the literature review. Major ideas will be addressed, while additional research will be presented more in the literature review chapter. The structure of this section follows firstly a presentation on investment in health, education, and social capital. The second part addresses measurement of health, education, and social capital. Finally, gaps in the literature—which will be addressed in this thesis—are presented.

The concepts of health, education, and social capital are not only nowadays used by economists, but are also employed by researchers in other disciplines such as sociology, health, and psychology. Nonetheless, these concepts are predominantly used by economists and are the fundamental concepts of growth and development theories. Despite this, there is a wide range of disagreement about the best approach for defining them. This thesis will employ these terms in a bid to better understand the growth and development process in the context of Vietnamese agriculture.

According to related economic literature, it is suggested that the definition of *investment* is essentially 'all expenditure which would not be made in a stationary state' (Scott, 1976). This means that investments include expenditures on new machinery and vehicles, buildings and constructions, on increasing stocks, research and development, and substantial expenditures on the marketing, planning and the education, health, and migration of people and their social relationships. Investment, therefore, is the cost of change, and subsequently covers all activities related to growth.

It must be emphasised that this thesis does not seek definitions of these concepts that will be suitable for all purposes. For example, the thesis does not address investment decisions. Project appraisal is effected using the discounted net cash flow approach, and expenditure is addressed with a distinction between 'current' and 'capital' transactions, and depreciation is strictly irrelevant. The thesis also does not address the efficiency of firms' investment projects. The thesis focuses on the measurements of investment of human health, education, and social capital in their relationship to income at the individual level. Moreover, the concern of the thesis is in the context of agricultural growth and development in a developing country. The fundamental idea of investment and capital is in this thesis addressed within a context of change.

Investment in health, education, and social capital

Investment in health is normally seen as consumption in health (Strauss & Thomas, 2008). Consumption is seen as an investment only when said consumption leads to an increase in productivity or income. Otherwise, consumption is simply to satisfy the demand of consumers for specific goods or services.

Investment in education is normally seen as years of schooling (Barro & Sala-i-Martin, 1995; Barro, 1997, 1999; Wolff, 2000, among others). Investment in education can be classified into two categories: (i) formal schooling and (ii) informal schooling. Formal schooling concerns the time workers spend on compulsory schooling at the early stages of their lives. Informal schooling refers to the time workers spend attaining further learning via their experiences or interactions with other people. An example of informal learning is presented in a research paper by Lucas and Benjamin (2011). In their paper, the learning process is defined as the time allocation and the obtaining of new knowledge through competition among competitors. However, this thesis focuses on formal schooling and applies investment in education as years of schooling.

In addition, investment in social capital is more complex than investment in education and health. Social capital is a multifaceted term that does not have firmly established definitions. Therefore, investment in social capital is usually composited from many sources such as: (i) investment in culture (Gelauff, 2003); (ii) investment in shared knowledge (Ostrom, 2000); (iii) investment in trust (Dasgupta, 2000; Fukuyama, 1999); (iv) investment in social networks (Ostrom, 2000; Slangen et al., 2004). The economic literature on investment in social capital presents not only these ways of thinking but other ways of understanding as well; however, it is not a realistic expectation for this thesis to cover all ways of thinking concerning this subject. Hence, the thesis analyses only the four above categories to frame the theoretical model and to analyse in the case of Vietnam.

Measurements of health, education, and social capital

Firstly, health and education capital are important because they have vital roles throughout the course of life. Health capital relates to the concept of death, while education capital relates to the concept of a better life. Therefore, in order to have a good strategy for better living, humans need to structure their investment plan so as to maintain good health and to increase the stock of their knowledge. It is therefore necessary to know how these types of capital are measured.

Parents are assumed to make key decisions for their children's health, whereas an adult is assumed to make his or her own decisions. It is important to distinguish between health outcomes and health inputs. Health outcomes relate to height, body mass index, disease incidence or physical functioning, while health inputs include nutrient intake, exercise, smoking, and utilisation of preventive or curative health care (Strauss & Thomas, 2008). The distinction between health outcomes and inputs is not always clear, as some health outcomes are intermediate inputs into other outputs.

For instance, disease incidence and body mass might affect one another. When we can see the distinction between health inputs and outputs, we can create the right strategy for investment in health. Health inputs will be objectives of investment in health; therefore the measurements of health inputs are important for investment in health.

The measurements discussed above are at the individual level; however, government policies stress these at the national level. In order to issue reasonable policies, the government uses analysis that employs aggregated data. At the aggregated level, health outcomes are, among others, life expectancy and disability-adjusted life years (DALY) (World Bank, 1990; WHO, 2000), whereas health inputs are government spending in health care, number of medical workers and hospitals, and government spending in medical research. Thus, government investment in health focuses on the utilisation of the current health care system and improvement of health facilities for its citizens.

Measurement of educational capital is similar to health capital. It has individual levels and aggregated levels. At the individual level, educational capital refers to a person's years of schooling or ability to learn. Normally, years of schooling, on-the-job training periods, or attending extra classes are used to describe educational capital at the individual level. At the aggregated level, school enrolment and adult literacy rates are employed.

Secondly, social capital is a multifaceted term and is hard to measure using only one variable. Practically, each dimension of social capital has its own unit. For example, Putnam (2000) indicates that social capital has five elements: (i) intensity of involvement in community and organisation life; (ii) public engagement; (iii) community and volunteering; (iv) informal sociability; (v) reported levels of interpersonal trust. Coleman (1988) believes that social capital has two types: (i) social capital inside family (ii) social capital outside family. For inside family, we have the relationship between parents and their children. For outside family, we have the relationship between parents and their children's school, relationships between relatives, and relationships among friends and social network members. Hence, there are numerous ways for calculating social capital, but unfortunately no common way to measure the specific *type* of capital. In short, there is currently no final conclusion for measuring social capital.

The most common measurements of social capital at the aggregated level are the performance and quality of institutions and the satisfaction and happiness of citizens regarding their government's actions. The effect of the quality of institutions on economic growth has been addressed by many researchers. Institutions considered to have high quality will boost economic growth and development, while low quality institutions will constrain economic growth and development. For instance, transformed countries in Eastern Europe and Asia display significant economic growth and development compared to periods prior to their transformation.

The gaps in economic literature

Firstly, the gap in economic literature related to investment in health is highlighted by Sachs and his colleagues (2001), who states that 'the importance of investing in health has been greatly underestimated, not only by analysts but also by developing countries' governments and the international donor community' (Sachs et al., 2001, p. 2), and 'although health is widely understood to be both a central goal and an important outcome of development, the importance of investing in health to promote economic development and poverty reduction has been much less appreciated' (Sachs et al., 2001, p. 1).

Secondly, this research aims to analyse the interaction between education investment and learning-by-doing (LBD) in the context of an education-income model for the case of Vietnam. Farmers in Vietnam do not have much education and enhance their knowledge in the fields only by working together. However, their foundational education plays a vital role in allowing them to obtain new knowledge from their co-workers, communication programmes of the Vietnamese government, and aid programmes supplied by international donors.

Thirdly, this research will explore the effect of investment on behaviour patterns to determine farmers' abilities to learn by doing under the analysis of the production function at the household and representative firm levels.

1.3.2. General analysis framework

In this thesis, I analyse the effect of investment in human and social capital on agricultural development at the household level in Vietnam in 2002, 2004, and 2006 from the perspective of socio-economic development, using an econometric method to model the link between farmers' income and their human and social capital. Moreover, the empirical results are used to suggest public policy to central and local government in Vietnam.

The rationale of this research is that the government implements a national targeted programme on poverty alleviation and sustainable development policy for rural development. The objective of the programme is

> 'to build a new countryside with gradually modern socio-economic infrastructure, rational economic structure and forms of production organization; to associate agriculture

with quick development of industries and services, and rural areas with urban development under planning; to assure a democratic and stable rural community deeply imbued with national cultural identity; to protect the eco-environment and maintain security and order; and to raise people's material and spiritual lives along the socialist orientation.'

This programme was started following Unification Day in 1975, but the methodology for implementing the programme has changed three times from 1975 to 2009.

- **1975–1980**: developing a new rural model at district level.
- **2000–2003**: piloting a new rural model at commune level (in 18 communes).
- **2007–2009**: community-based new rural model at village level.

In the context of implementing a new method for rural development, this research aims to investigate the preparation of villagers for exploiting the opportunities established by new government policy.

The expectation I have is that to some extent, farmers invest their human and social capital in order to gain more income in the future. Additionally, the link between generations of farmers, proxied as the investment of parents to their children, expresses parents' care about their children. Farmers in Vietnam are usually considered to be poor households and their investment in their children supports the expectation that the next generation will be equipped with adequate knowledge and skills for working in the industrial sector, or to escape from agricultural labour. The return of human and social capitals on agricultural household incomes can be seen as a good source of inputs for implementing a vital programme with which to sustainably alleviate poverty in rural and agricultural areas. The investment of farmers in human and social capital depends on the rate of return; if the rate of return is high, farmers will invest more in human and social capital rather than saving their money in banks.

At the provincial level, social capital is represented by the performance of the provincial government. The performance of a provincial authority, in turn, is measured by the businessmen who run their businesses in the province. Variables are conducted from a set of dimensions that concerns human capital development policy, the legal system, or state-owned enterprise-biased policy. This analysis is more complicated than the other types of capital and will be discussed comprehensively in chapter 7.

To measure the rate of return of investment in health, education, and the effect of investment in social capital on the incomes of agricultural households, and as representative of agricultural enterprises' revenue at the provincial level, the general analysis framework of the thesis begins by addressing production function. The function is stated in the Cobb-Douglas format—*labour-augmenting* or *Harrod-neutral* type—as follows:

$$Y_t = f(K_t, A_t L_t) \qquad (1.1)$$

where at the time t, Y is gross domestic product, K is physical capital, L is labour and A is 'knowledge' or 'effectiveness of labour'; therefore, AL is effective labour. In economic literature, the effects of K and L on Y are obvious, while the effect of A on Y remains a mystery and of interest to many researchers. Syverson (2011) offers a comprehensive analysis of the subject.

According to the scope of endogenous growth theory, 'knowledge' or 'effectiveness of labour', A, is a function of learning-by-doing *(LBD)*— knowledge and skills obtained from working and from modern technology. The function has the following form:

$$A_t = f\left(LBD_t, A_t^*\right) \quad (1.2)$$

For a simple analysis consistent with the circumstances of Vietnamese agriculture, its general meaning does not lose when assuming that A_t^* is a constant. This means that all farmers have the same technological endowment and their different productivities only depend on their methods of harvesting and obtaining experience through working together. Therefore, function 1.2 could be expressed in the next function as

$$A_t = f(LBD_t) \quad (1.3)$$

Substituting function 1.3 to function 1.1 yields a general production function:

$$Y_t = f(K_t, LBD_t, L_t) \quad (1.4)$$

Dividing both sides of function 1.4 by L_t, we have

$$y_t = f(k_t, LBD_t) \quad (1.5)$$

where at time t, y is gross domestic product per capita, k is physical capital per capita, and *LBD* is labourers' knowledge and skills obtained from working. Function 1.5 is the starting point for further analysis in this thesis.

The *LBD* capabilities of an individual are affected by his/her health, educational attainment, and social capital or 'behaviour pattern' (Solow, 1999).[4] The relationship between them can be expressed as the following functions:

$$LBD_t = g(health_t, education_t, \text{'behavior pattern'}_t)$$

Or

$$LBD_t = g\,(health_t, edu_t, bp_t) \quad (1.6)$$

To see the relationships between learning by doing, health, education, and behaviour pattern, it is necessary to understand the definitions of these terms, as these themselves point out the relationships. A Nobel laureate in economics, Professor Kenneth Arrow, defines learning as a product of experience and also promotes a definition of learning by doing. Arrow (1962, p. 155) states, 'Learning can only take place through the attempt to solve a problem and therefore only takes place during activity.' This means that an individual cannot learn without interacting with other people, because solving a problem requires communication with others, no matter what form this communication may take. In order to have effective communication, people are required to have at least a basic educational

---

[4] In the thesis, the terms 'social capital' and 'behavior pattern' are used interchangeably.

background and be in good health. A basic education helps people to read and understand basic knowledge; it is a bridge for people to move into higher education or deeper knowledge. Good health conditions enable people to maintain the ability to learn and work.

Education is a major part of the learning process. Education can be defined as 'an institution specializing in the production of training, as distinct from a firm that offers training in conjunction with the production of goods' (Becker, 1993, p. 51). Formal school provides a wide range of general knowledge, while further learning specialises people into industries. Therefore, education affects the learning-by-doing performance of farmers in their later lives. Farmers who have had better education will usually have better chances of learning more during their agricultural activities, while farmers who have had poor education might not have the same opportunities for gaining advanced knowledge or skills.

Health is defined as a state of complete mental, physical, and social well-being and not merely the absence of disease (WHO, 1949). The measurements of health vary, because 'health' is a multi-dimensional combination of an array of factors. These include mortality, morbidity reports, health-related behaviours, self-assessments of overall health, assessments of physical functioning or activities of daily living, information on specific morbidities, and a battery of physical assessments that include biomarkers. Healthier people tend to be more economically productive, energetic, and mentally alert. A healthier farmer can harvest more crops, while a healthy worker can increase cognitive ability by learning new skills. Healthy people can expect to live longer; they therefore have greater incentives to make longer investments in both their farms and human capital.

In the literature of political science, sociology, and anthropology, *social capital* generally refers to the set of norms and organisations through which people gain access to power and resources that are instrumental in enabling decision-making and policy formulation. Economists add to this focus the contribution of social capital to economic growth. At the microeconomic level, they view social capital primarily in terms of its ability to improve market functioning. At the macroeconomic level, they consider how institutions, legal frameworks, and the government's role in the organisation of production affect macroeconomic performance. The complexity of social capital will be addressed further in the literature review chapter.

Substituting function 1.6 for function 1.5 yields

$$y_t = f(k_t, g\,(health_t, edu_t, bp_t)) \quad (1.7)$$

Health, education, and *'behaviour pattern'* might have correlations, as addressed in current related economic literature. For example, Goldman and Smith (2002) suggest that better educated individuals are better at self-managing health problems by adhering more closely to treatment protocols. The interaction between these variables will affect the outcome of estimation. Thus, to avoid an unreliable estimation, function 1.7 has been broken down into the following three functions for estimation purposes:

$$y_t = f(k_t, g\,(health_t)) \quad (1.7.1)$$

$$y_t = f(k_t, g\,(edu_t)) \quad (1.7.2)$$

$$y_t = f(k_t, g(bp_t)) \quad (1.7.3)$$

Estimations of the three above functions will be done to address a set of research questions:

1. To what extent does an individual's human and social capital affect farmers' income in the context of Vietnamese agriculture?
2. Could consumption in *health* and maintaining *'behaviour pattern'* be seen as investments?
3. What is the possibility that further public policies on agricultural development will be derived from results of the analysis?

1.3.3. Findings and public policy implications

The empirical results of this thesis explore the effect of investment in health, education, and social capital on agricultural households' income. In this section, the most significant results are presented; other results are presented in the appropriate chapters.

Firstly, as expected, consumption in health had a positive effect on agricultural households' income. The positive relationship between the increase in health consumption of households and the growth of the households' incomes implies that consumption in health in agriculture at the household level in Vietnam has had an investment effect. Therefore, the rising consumption in health might lead to growth in household incomes. However, the effect of investment in health on agricultural households' income in Vietnam remains low (for further details, see chapter 5). It leads in this instance to the policy implication that the government either should or should not support agricultural households in investing in their health. If the government agrees to support them, then the next question concerns the implementation procedure of these policies. Based on the empirical results from this analysis, the only answer that can be given in this instance is that the government should support agricultural households in investing in their health. For the second question, from the outcomes of the analysis, some logical suggestions could be provided to government officials. Additionally, another interesting result was that, constrained by a lack of health investment, farmers were relying on their vintage knowledge at a rate of just above 80 percent between 2002 and 2004.

Secondly, investment in education, proxied by years of schooling, has positively and significantly affected agricultural households' income. Additionally, the high value of the rate of return of investment in education implies that under-investment is occurring in Vietnam's agricultural sector. This result is new in the case of Vietnam, where many other articles indicate an over-investment situation. Using Hausman and Taylor's regression method and considering individuals' unobservable influence, the coefficient of years of schooling was dramatically increased, from nearly 3 percent to more than 20 percent between 2002 and 2006. The increase between the two coefficients indicates that unobservable individual ability considerably affects the capacity for learning of the individual farmer. The outcome implies a public policy for supporting further learning of farmers. Improving the quality of current vocational training schools can also be seen as a significant area of support from the government to farmers.

Thirdly, investment in social capital has a considerable effect on agricultural households' income at the household level, while at the provincial level, the effect of the quality of local government on a representative firm's outcome remains unclear. As stated in the literature review, it is hard to capture social capital into one variable. In this thesis, two sets of variables at the household and provincial level were selected to analyse the effect of investment in social capital on agricultural households' income, as well as on agricultural firms' revenue. At the household level, the sign of coefficients indicated that investment in the social capital of farmers in 2002, 2004, and 2006 supported the economic theory on investment in social capital. At the provincial level, the signs did not entirely support the economic theory, due to a lack of data and the imperfect matching of two datasets.

Based on the empirical results of the relationship between investment in social capital and economic growth at household and provincial levels, the improvement of quality public policy implementation process by local government should be seriously addressed. Additionally, the government should focus on the cohesion of family members, social ties, and social infrastructure. However, at household levels, the most important public policy the government should emphasise is increase in household income.

1.4. The contributions of the study

This dissertation aims to contribute to the current economic literature concerning the role of investment in health, education, and social capital on the productivity of farmers in Vietnam. These types of households were selected for three reasons. *Firstly*, they play a key role in the early stage of the *Doi moi* strategy. *Secondly*, it has been widely observed that in Vietnamese agriculture, a large number of agricultural households are using obsolete and inefficient machinery. This indicates that any increase in agricultural productivity in Vietnam will be constrained. Therefore, if Vietnam wishes to promote further development in agriculture, the government needs to provide strong support to agriculture. *Thirdly*, the Vietnamese government is implementing a new rural development programme, so there is a need for analysing the current situation of agricultural households' capitals. The core element of the new programme is that the government wants its farmers to independently decide on suitable investment activities based on their needs and current assets. Thus, the three research questions have been raised; by providing answers to them, the contributions of this thesis are six-fold.

Firstly, this thesis is the first trial to use an income-health model by employing the VHLSS datasets to measure the rate of return of investment in health on income at the household level with a focus on Vietnamese farmers. In Vietnam, research using the VHLSS datasets is not rare; however, studies on the incomes of agricultural households are. Almost all existing research uses household expenditures, proxied for households' well-being as the independent variable, and almost all existing studies focus on explaining the pattern of households' well-being. The major reason for this is that household incomes are difficult to estimate. The difficulty of household incomes estimation stems from unreliable data sources. The estimation of agricultural household incomes is more difficult than non-agricultural households, because their major source of income is self-supplied.

Additionally, studies in health usually emphasise health status and health indicators, while few studies use the income-health model to analyse the effect of health investment on income at the household level. Therefore, this thesis aims to fill the gap in related economic literature by estimating the relationship between households' income and their investment in health.

Secondly, the thesis measures the rate of return of investment in education on agricultural household incomes in Vietnam using Hausman and Taylor's method. This method requires panel data; however, the VHLSS datasets are not really panel data. Only some households are re-interviewed during the data collecting waves. As a result, not many researchers choose this method to measure the rate of return of investment in education. Nonetheless, Hausman and Taylor's method can estimate the unobservable individual effect, which indicates the ability of farmers to learn from each other. Unobservable individual effect estimation can help estimate the learning-by-doing of farmers. In fact, using Hausman and Taylor's method will lead to an increase in the education coefficient, but the reason behind the increase is not usually well explained. The author has established that using this method can describe the true facts of investment in education in Vietnam, because the empirical results indicate a situation where under-investment is occurring in Vietnam's agricultural sector.

Thirdly, the thesis measures the effect of maintenance of inter-family relationships on agricultural household incomes. These relationships are very important in the cultural context of Vietnam. It is traditional to the development of rural society, but has seen a decrease since the implementation of the *Doi moi* policy. It is difficult to explain the reasons for the decrease, yet it is clear that in the development of rural economics, the relationships between rural family members are changing. This change might affect the increase of agricultural household incomes because of the rising mobility status of families. It is too soon to say whether the change will be positive in the long term or not. However, this change does not harm the development of rural families. In order to capture the effect of the adaptation on the productivity of agricultural households, proxied by income, this thesis wishes to contribute to the current economic literature by estimating this effect. It was found that the evaporation of family relationships does not constrain the development of household incomes. The finding is in keeping with results from other studies in other countries and communities worldwide.

Fourthly, the thesis is the first trial to measure the effect of time use on social bonding relationships and how this affects agricultural households' income. The allocation of time is a current debate in economic literature and the effect of this allocation on income has to date not received much attention at the household level, in all likelihood because it is not an easy task to accomplish. Time is unobservable; therefore, the allocation of time is difficult to estimate. As a result, estimating the effect of allocation of time on income is a risky task. The outcome of such estimation may not be reliable due to a lack of appropriate data. In this study, the author struggled with this situation as well. To overcome this difficulty, a simple method was applied. Using information regarding time spent working and with an assumption about resting time, the thesis states that for the rest of the day, time will be allocated to the maintenance of relationships between people. This can be seen as a good way for estimating the effect of the binding of social relationships on agricultural households' income; however, there is nonetheless the need to be very careful in interpreting the results. Regardless, the empirical results support the idea that binding social relationships might help farmers raise their income. Therefore, a policy implication in this instance is that the government should support farmers in establishing a place for organising where they can meet with one another. It also highlights the reasons for why so many community halls have been built in Vietnam. Additionally, this result also supports the idea that a criterion for a new rural development programme in Vietnam is for all communes to have their own community halls.

Fifthly, the thesis provides a first analysis of the relationship between the performance of local government and the revenue of agricultural firms at the provincial level for the case of Vietnam. Almost all studies on firms employ data at the firm level and analyse the effect of trade policy on the revenue of firms; less research is concerned with the effect of the quality of local government on the revenue of firms. This situation is due to a lack of relevant data. Fortunately, since 2005, USAID has conducted a survey measuring the quality of local government via firms' evaluation. The survey involves the institution (USAID) asking firms in all provinces in Vietnam to evaluate the performance of local government. The outcomes of the survey are published annually. The availability of this dataset has allowed the author to construct a panel of data at the provincial level for measuring the effect of the performance of local government on agricultural firms' revenue. However, the empirical results of this thesis will still need further expansion in the future.

Finally, the thesis serves as a comprehensive suggestion on public policy frameworks to the Vietnamese government for further agricultural-rural development, as based on the empirical results of the study. The new rural development programme has been in place since 2010, but the preliminary analysis was done over the span of five years, from 2004 to 2009. The analysis, however, served only the purpose of making the decision that the programme needed to be implemented. The future of the programme's implementation is therefore not systematic. In order to advise the government on further policy implementation, this thesis builds up a policy framework based on its empirical results. Policy implications cover three main areas of the programme: rural health, rural education, and rural community.

1.5. The structure of the thesis

After the introduction chapter, chapter 2 focuses on the historical context of the subject and the current public policy debate regarding the new model of socio-economic development of rural and agricultural areas in Vietnam. In order to highlight the position of rural and agricultural development in Vietnam among common worldwide trends, this chapter also addresses the current trend of world agricultural development, which is summarised in the recent report of the World Bank titled 'Agriculture for Development' (2008).

To set the theoretical and empirical framework for the research, chapter 4 provides an in-depth analysis of current economic literature on human and social capital and the relationship between the two types of capital and agricultural development. Relevant economic literature covers the scope of theory from the micro-level to the macro aspect. The scope includes several definitions of human and social capital and a way for measuring the effect of investment in human and social capital on agricultural development.

Chapter 5 addresses the relationship between individual health investment and income within the framework of an income-health model. This chapter explores the data on health and income from the VHLSS surveys in 2002 and 2004 and data on life expectancy at the provincial level. The selection is due to the fact that the VHLSS does not have a rich set of health indicators and the health survey does not have any information on individual income. The empirical results suggest that individual health investment has positively affected income. This result suits the current socio-economic development theory.

The second aspect of human capital investment is investment in education. The link between education and income is analysed in chapter 6 using the income-education model. Using panel data from 2002, 2004, and 2006, this chapter discovers the robust rate of return of investment in education on income. Employing Hausman and Taylor's regression method, with controlled individual unobservable effect, the empirical results of the chapter implies that the government should provide better educational facilities for rural people in order to boost the rate of income growth in rural areas.

Chapter 7 focuses on how social capital investments of individuals affect their income. This chapter draws on data from a range of sources, including the VHLSS, agricultural firms via the General Statistics Office, and data on the performance of provincial governments via USAID. The empirical results confirm that the current policy of local government is limiting the development of agricultural households and firms. The implication, therefore, is that central and local government should effect further reformation in order to achieve their goals for the new rural development programme.

Following this, chapter 8 presents a wide-ranging framework for public policy on the socio-economic development of agriculture. The framework addresses three important activities of the current New Rural Development

Programme (hereafter NRD). Moreover, the chapter suggests that local and central governments should consider the findings from this research in order to reach the goal of NRD by 2020.

A concluding chapter presents a summary of the findings of this research and proposes further studies for the future.

Chapter 2

HISTORICAL CONTEXT AND POLICY DEBATES

2.1. Introduction

Agricultural development has long been at the centre of dialogue in policy debates in Vietnam. Recently, with the new policy and a new rural development strategy, rural and agricultural development has again been pushed to the forefront for discussion. During the War of Independence (1945–1954), the anti-colonial resistance movement, Viet Minh, sprang up in the agricultural sector with the appropriate land policy. Agricultural land had been transferred to farmers from the French or Vietnamese, the latter having been supported by the French (M. Ravallion & D. Van De Walle, 2008). In the North, this policy was applied to different types of farmers in order to forcibly remove rich peasants from positions of power. This policy made farmers equal in terms of economic and political power. Subsequently, around 1957, collectives were employed in agricultural development. It was seen that under the effects of the policy, the inequalities and class divisions that had plagued Vietnam since French colonisation had been redressed and prevented.

Following the French defeat, consecutive US-supported governments put a premium on agricultural development with a focus on land issues; however, the government had concentrated on landlords with large areas of land rather than small and medium farmers (Callison, 1983). At the same time, the resistance movement led by the National Liberation Front (NLF) emphasised land-rent reductions and redistribution to small farmers or landless households who were living in the controlled areas. The Saigon government realised that the policy of NLF might attract the support of their people, so they introduced the Land-to-the-Tiller (LTT) programme (Callison, 1983; Wiegersma, 1988). This programme provided cultivators the ownership rights through land titling and restricted the size of landholdings to no more than 20 hectares. However, the programme did not have the impact desired by the government, and soon after the withdrawal of the US and the country's re-unification in 1975, this policy was abandoned. The large tracts of land were redistributed and collectivisation was applied to the whole of Vietnam. However, this policy of the Vietnamese government was largely unsuccessful because of the resistance on behalf of farmers.

Under the cooperatives policy in the North, peasants were formed in production brigades ranging from 40 to 100 and run by the brigades' heads. The head signed a contract with the cooperative to supply outputs to the cooperative, assigned the work to brigade members and collected reports from the members. The performance of farmers was measured by days of work, from 200 to 250 days per year. Payment was in output units such as paddy or corn, based on personal labour contribution. In the South, farmers were forced to join 'collectives' in order for the whole of country to have collectivisation. Farmers in the South could cultivate privately on their temporarily assigned land, but their tools were shared and all their inputs and outputs managed collectively. In short, collectivised agriculture had been formed in 'cooperatives' in the North and 'collectives' in the South.

The cooperatives had made sense in the North during the War of Independence, but made less sense to agricultural development following the reunification of Vietnam in 1975. Farmers had widely resisted the policy on cooperatives, and 'villagers' everyday politics gnawed the underpinnings of the collectives until they collapsed. Rural households, for the most part, wanted to farm separately' (Benedict J.Tria Kerkvliet, 2006, p. 285). In truth, peasants had private land at the beginning of collectivisation, with total private land not over 5 percent of collectives' land; nonetheless, some of the areas might have been larger than 5 percent. Peasants could use their land for growing vegetables or other produce not supplied by the cooperatives. Interestingly, the productivity on this land was normally higher than on collective land in terms of monetary value. In certain areas, localities ran a surreptitious production system called a 'sneaky contract' (or 'khoan chui'), leading to the leaders in these areas being accused law violation and forced to retire. Consequently, the collectivisation system imploded.

During the '70s and '80s, many Vietnamese rulers and urban elites were unhappy about collectivisation. They accused collectivised farming in the late 1970s to be the major reason for the reduction of agricultural productivity. Food shortages were common during this period and the government faced a multitude of pressures. The War of Independence had been costly and destructive, and had left in its wake many injured and displaced people. These problems coincided with the dreadful economic situation of the late '70s. Correspondingly, the centrally planned heavy industrialisation had performed poorly and the deterioration of the relationship with China led to the reduction of food aid from that country. The border war with China in 1979 was seen as a final attempt to protect ideology of the doctrinaire old guard of the Communist Party, before losing the debate with pro-market reformers in the party.

A clue to a rethinking of agricultural development policy might be found in a party document form 1979 (Benedict J.Tria Kerkvliet, 2006). In the early '80s, several adjusted policies on agricultural development had been implemented, such as Contract 100, which allowed localities to sign a specific production quota contract with agricultural households instead of signing a work contract as before. In 1986, the Sixth Congress of the Vietnamese Communist Party announced the retirement of old leaders and their replacement with a number of well-known reformers, such as the Former Communist Party Secretary Nguyen-Van-Linh and Prime Minister Vo-Van-Kiet. *Doi moi* and other reforms followed the strategic adjustment of the party, which will be recorded and analysed in the following sections.

The structure of this chapter is as follows. The first section offers a short introduction. The second section presents a summary of global agricultural development in order to understand the position of Vietnamese agriculture in the historical flow of agricultural development. The third section describes some major historical points of agricultural development in Vietnam. The last section presents the policy debate related to agricultural development and a new direction for agricultural development in Vietnam.

2.2. Stages of agricultural development worldwide

* **The proportion of rural population is high in developing countries, particularly in most Asian nations.**

Overall, according to a recent report published by The Worldbank,[5] global rural population size has declined since 1980. In 1980, rural populations made up 60 percent of the world's total population. This number reduced to just over 50 percent in 2006. This means that half of the world's population is agricultural. In low-income countries, the proportion of rural population decreased from 78 percent in 1980 to nearly 70 percent in 2006, while in middle-income countries, the process appears inversed; the proportion increased from 44 percent in 1980 to just over 45 percent in 2006, after peaking at 56 percent in 1990. In addition, East Asia and Pacific countries have had the most significant adjustment in rural population percentage, while South and Central Asian countries have changed slightly. The percentage was reduced by 20.4 percent in East Asian and Pacific countries, whereas South and Central Asian countries have had a lower rate of changing the percentage of rural population, 6.8 and 4.8 percent, respectively. In Latin, Caribbean and Sub-Saharan countries, the percentage went down moderately from 35 and 77 percent in 1980 to 22.3 and 64.2 percent, respectively. The percentage in Middle Eastern and North African countries decreased considerably, from 52 to 42.5 percent in the period 1980–2006.

In Asia, the trend of a declining percentage of rural population was identical to the common global tendency. Indonesia experienced the largest

---

[5] For detailed statistics, please see WorldBank (2001, 2007b).

change, followed by the Philippines, Malaysia, South Korea, China, and Thailand. The percentage of Indonesia reduced from 78 percent in 1980 to 50.8 percent in 2006. In the Philippines, Malaysia, South Korea, China, and Thailand, rural population decreased by 26.4, 26.2, 24, 23.3, and 15.6 percent, respectively. This trend reflects the rate of urbanisation, industrialisation, and economic growth in these countries. The second group had the lower rate of transformation and ranged from 4 to 11.5 percent. Surprisingly, Japan and Sri Lanka increased their percentage of rural population in the period 1980–2006. In Japan, the percentage rose from 24 percent in 1980 to 34 percent in 2006. In Sri Lanka, this percentage was raised by 6.9 percent at the same period of time. More detailed information is shown in the table below.

This transformation process was due to organisational and institutional changes, as well as scientific advances. Farmers have contributed the majority of dynamics for this process. This development was possible because farmers were able to produce more than they consumed and could export their products to urban areas.

* In the developing countries, agricultural growth plays an important role in economic growth.

In the long term, world agricultural production is higher than the rate of population growth. According to a World Bank report in 2008, from 1980 to 2004, developing countries had a higher rate of agricultural growth than developed countries, 2.6 percent per year in comparison with 0.9 percent per year. In addition, the share of developing countries' agricultural GDP has risen from 56 percent to 65 percent, far higher than non-agricultural countries. This situation is due to the high productivity of

Asian agricultural countries. Globally, low-income countries tend to have a higher proportion of the labour force in agriculture, and this portion will be reduced when their incomes are raised. In the low development countries, rural labour is around 60 percent and sometimes as much as 80–90 percent. However, in particular countries such as Nigeria and Brazil, the shares of agricultural labours are reduced; this is, however, not due to a higher GDP per capita. Instead, this problem may a result of the urbanisation process, which leads to agricultural labourers migrating to urban areas to seek better jobs. Moreover, China's case of economic development also supports this unusual situation.

* The stage of agricultural development is presented through the transformation of agricultural markets.

Based on the literature and history of agricultural development worldwide presented by Reardon and Peter (2007), the development of agriculture in developing countries can be focused on the transformation of markets for agricultural output. From the authors' analyses, the transformation process has five phases that can be categorised in two stages:

- Stage 1 (1950s–1990s): priorities in domestic development with four phases: (i) 1950s to 1960s; (ii) 1970s to early 1980s; (iii) early-mid 1980s to early-mid 1990s; (iv) 1990s.
- Stage 2 (mid 1990s–present): extreme globalisation of agricultural products in developing countries.

In the first phase (1950s–1960s), most developing countries became independent following extended periods as colonies of other, more developed countries such as the United Kingdom, France, Spain, and Portugal. During this time, farmers could not produce large commodity surpluses (Lewis, 1954). In most colonised countries, governments supplied enough and cheap food for their people. As a result, their policy focused mainly on the following:

(i) Physical infrastructure investment
(ii) Taxes on agricultural tradable outputs to provide revenues for the governments and permit physical infrastructure development

The second phase began in the 1970s and ended in the early 1980s. This phase emphasised improving income distribution of the economy. In this period, the Green Revolution started in Asia and diffused to the other continents. Some Asian countries and Latin American nations were well-developed in terms of economic performance and ready to transform to the second phase, while others remained in the first phase, their policy remaining in the earlier period. This situation provided an opportunity for lower developed countries to add new policies in addition to their old priority strategies as a result of the opportunity to apply Green Revolution technology. Consequently, new policies were improved owing to some positive aspects:

(i) Cheap global credit was available for every country in the 1970s
(ii) Capital, fertiliser, seed, and machinery were subsidised
(iii) To stabilise the process of agricultural development, the demand side was subsidised and the supply side was supported

In the third phase, the purpose of transformation of agricultural markets aimed at supporting the early stages of industrialisation in developing countries that had been implemented since the 1950s to the late 1970s. The earliest change happened in Asian countries, while later reform happened in African nations. The key factors of changing policy during this period were as follows:

(i) Cheap global credit was not abundant, and available only in small and medium scale after the mid-1980s
(ii) Inputs subsidised were no longer available
(iii) Subsidised demand side and supported supply side were phased out
(iv) GATT and later WTO became the main legal system for international markets for agricultural outputs
(v) The deregulation in foreign direct investment in agriculture was implemented

A decade later in the 1990s, phase 4 was implemented with the main achievement of institutional improvement. Phase 4 focused on fixing market failures through justification of institutional reform. To correct market failures, the government replaced price policies with non-price polices, such as supporting non-government organisations (NGOs) to assist farmers who could not obtain benefits from the previous stages. Moreover, NGOs supported institutional and legal enhancement subject to the reduction of transaction cost. This progress was presented in the World Bank's report, titled *World Development Report 2002: Building Institutions for Markets.*

The fifth phase started after the mid-1990s under the globalisation process. In this period, market liberalisation and foreign direct investment can be seen as two major pillars for agricultural transformation. Market liberalisation was addressed by the changing of market structure and the raising of the supermarket system. Market structure was transformed through structural adjustment programmes or by multilateral agreements such as GATT, then WTO, MERCOSUR, and NAFTA, to which countries subscribed. Moreover, the market structure was amended by transportation and storage improvements, and government investment in infrastructure and tax credits. Supermarkets were originally seen as a place for the rich to buy supplies from, but this perception was expanded to include other classes of consumers. Agricultural transformation had led to a rise in farmers' incomes, which led to an increase in the demand for supermarket services. In addition, investment of supermarket firms helped to raise and meet the demand of farmers in terms of supermarket services. On the demand side, the expanding of supermarkets was pushed by large-scale processing industries and the growth of public transport and car ownership. Consequently, the prices of processed products were reduced.

Additionally, the rate of agricultural transformation was mechanised by foreign direct investment. The major part of foreign direct investment is the food industry, because it has a much higher rate of return than primary production. The presence of foreign direct investment in processing and retail industries can be seen as evidence for the status of each country adapting to globalisation and market liberalisation. Moreover, with the rise of the supermarket system, globalisation, in general, and foreign direct investment, in particular, has affected agricultural transformation in developing countries. This process sets domestic prices of agricultural output closer to international prices.

2.3. Actions for agricultural development in Vietnam

2.3.1. De-collectivisation

Under the Land Law of 1988 and its directive, Resolution 10, the members of collectives granted the individual long-term use rights to own agricultural land. The implementation of the law had begun the process of de-collectivisation, a process that was largely completed by 1990 (Beresford, 1990; Long, 1993). Resolution 10 made many suggestions; under the new law, collectives signed a contract with their members for using collectives' land over a period of 15 years (the time of using land is up to 50 years due to certain perennial crops). This policy encouraged peasants to invest in land improvement. For the first time in modern history, farmers were free to sign a service contract with collectives or individuals and could hire seasonal or employ permanent employees based on negotiation. Cooperatives held only some of the important production means that all members could access, such as tractors, and sold all other materials to their members. The contracts between cooperatives and farmers were in the past based on market prices instead of fixed prices. Payments were derived from piecework rates rather than work-point, which was a principal way of paying under the collective regime.

Under the combination of the political system of central authorities and decentralised localities, villagers took part in decision making and self-management. All villagers had equal opportunities for gaining political promotion and having access to power, which helped to advance revolution at the grassroots level. The de-collectivisation process enabled local cadres or their relatives to accumulate more agricultural land than other peasants. It appeared to be very difficult to effect land redistribution without corruption. Serious disputes occurred in the Mekong River Delta, with the most extreme case being the march of farmers in Ho Chi Minh City (Beresford, 1990). The major cause of the protest was that cadres had used their rights to distribute to their family the lion's share of land. In the North, in the Hai Minh commune (Ha Nam Ninh Province), only 60 percent of cooperatives' land had been distributed equally; the remaining 40 percent had been allocated to agricultural households who had capital and experience in rice production (Beresford, 1990). Hai Minh commune cadres used complicated and tricky classifications in terms of capital and experience, which led to poor farmers not having capital and experience in rice production, while middle class and rich farmers were the relatives of the commune's authorities. The differential distribution of collectives' land had led to the differential income of local agricultural households and had brought about the dissatisfaction of farmers on the implementation of Resolution 10.

The contracted product system was introduced in the North of Vietnam during 1981–1982 and in the South of Vietnam during 1982–1983. This was an attempt to improve the performance of cooperatives in the North and collectives in the South; however, this solution caused the reversion of private property, which was restricted after the Reunification. Initially, only labour tasks were contracted with households, but during the latest wave of reform all production processes had been contracted directly with households. Individual households had sole responsibility for the productive outcome of labours and their incomes depended directly on this outcome. In the latest stage of reform, all cooperatives' assets had been sold to individual households. The objective in this instance was that the Vietnamese government wanted to force the pace of abolishing subsistence farming and running a commodity economy for the whole country. This goal was suitable with the suggestion of a modern development economics discipline where, for developing and overpopulated countries like Vietnam, it is preferable that agricultural development for small-scale farmers be developed instead of for large-scale farmers (Lewis, 1966).

In order to improve the performance of collectives, the Vietnamese Communist Party adjusted the style of collectives in the mid-1990s in terms of structures and ways of formulating collectives, and called the new style 'New Style Collectives'. This movement was initially based on two field expeditions in the Thai Binh and Thanh Hoa provinces. The two field expeditions had led to a provincial conference titled 'Transformation of the old-style cooperatives and the development of new cooperative forms in the rural areas of Thanh Hoa' (April 1995), which focused on the real situation of collectives in Thanh Hoa. During the conference, a party of opinions had been served for participants with a wide range of ideas, including the idea that the dissolution of old-style cooperatives would destroy socialism. Following this, the provincial party organisation issued Decision 9/NQTU on the transformation of agricultural cooperatives and the implementation of new collective forms in Thanh Hoa Province. In September 1995, the People's Committee of Thanh Hoa Province issued a decree to guide the implementation of Decision 9. The success of the experience in Thanh Hoa and Thai Binh provinces inspired confidence in the Vietnamese Communist

Party that the new style cooperatives could solve the problem of the old style cooperatives. The new style of cooperatives was then applied in the transformation progress of cooperatives and called an Agricultural *Service* Cooperative.

In short, de-collectivisation in Vietnam is actually a means of forming collectives in a top-down to down-top process, from the ideas of leaders to the needs of farmers. The democratic approach of forming collectives has proven that when collectives are built based on the needs of farmers, they will be managed efficiently.

2.3.2. New Policies on agricultural development

In recent years, under the *Doi moi* policy, agricultural and rural economies have achieved significant contributions to food security, export growth, employment generation, poverty alleviation, and social equality. However, expected goals and actual results have not entirely matched. Major problems still exist because of the slow transformation of the economic structure, fluctuating economic growth and fragmentation of agricultural production. Recently, agricultural and rural areas have attracted nearly 50 percent of social labour and more than 70 percent of the population. This situation put pressure on the government regarding the improvement of living conditions and standards for rural and agricultural households.

Figure 2.1: The density of rural populations of Vietnam from 1930 to 2010

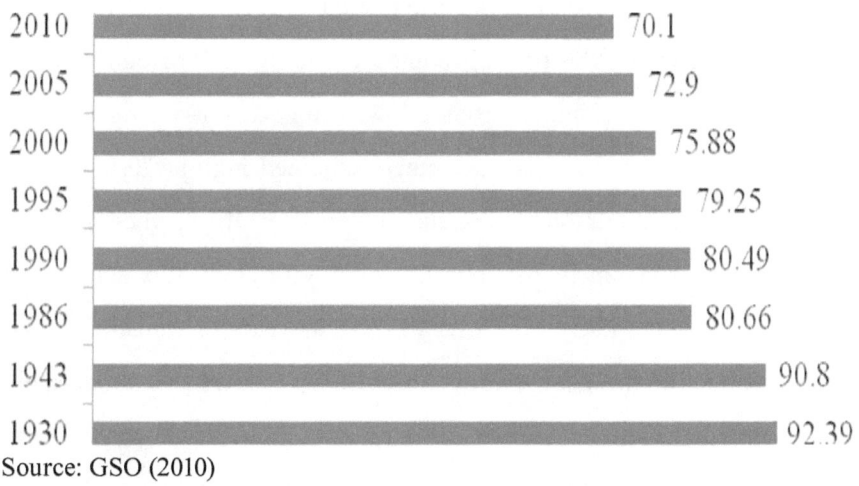

Source: GSO (2010)

Figure 2.2: The density of rural labour of Vietnam from 1990 to 2010

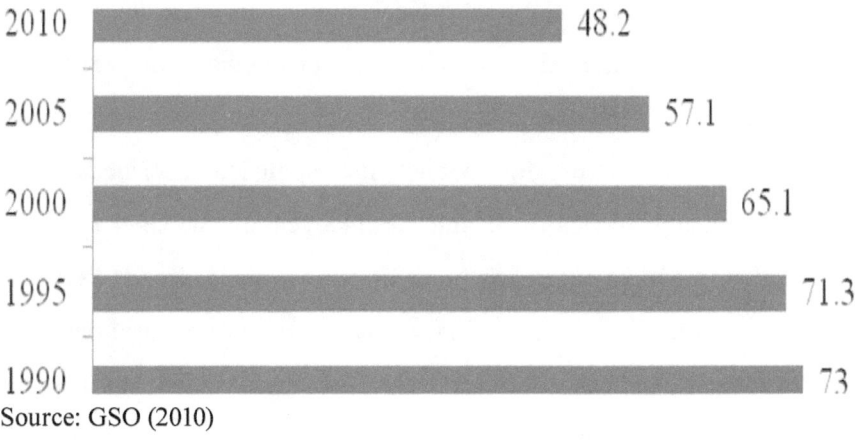

Source: GSO (2010)

In 1930, the rural population was 92.39 percent, and two years before the August Revolution in 1945, the percentage of the rural population was still high at 90.8 percent. In 1986, the proportion of rural population was 80.66 percent, while agriculture and rural development were seen as top priorities in the economic development strategy of the government. At the end of the economic crisis in 1993, the percentage had decreased slightly in comparison with 1986 (the difference was only 0.71 percent). When Vietnam's economy changed from a low-income country to a middle-income country in 2010, this percentage remained high at 70.1 percent. According to the estimation of the General Statistical Office, when Vietnam essentially becomes an industrialised economy in 2020, this percentage will be 60 percent. Therefore, after thirty-four years of transformation, the percentage of the rural population will have decreased moderately by 20.66 percent. This result is still high within the ASEAN context, and places significant pressure on the government's social policies related to education, healthcare, job creation, income raising, living standards, and other environmental issues (WorldBank, 2007a).

Despite numerous investment projects in infrastructure development implemented by central and local governments, the technical infrastructure for rural development remains poor and non-synchronous. The current infrastructure does not help the expansion of services, trade enlargement, and agricultural and rural development. Inter-regional infrastructure and services are weak or not available. Roads between villages, communes, and from villages and communes to rural business centres are not good enough to serve the business activities of villagers; this includes water supply and drainage, electricity supply, and lighting and water shortages and/or water pollution. Uncontrolled waste water and garbage disposal is harmful to people's health. Thus far, in many rural areas, infrastructure projects are serving shorter goals and there is a lack of long-term infrastructure investment projects, leading to the performance of rural infrastructure becoming out of date quickly during the development process.

To develop rural areas, the government realises that they are faced with many constraints and challenges. They also know that the rural transformation strategy might not be promptly promoted due to the following issues:

- Most leaders' awareness of the content and methodologies for the new rural areas' development remain limited (management by administrative orders is still dominant, there is a gap between the government policies/direction and lower level agencies' acquisition capacity and responsibilities for implementation, etc.). A number of people and local staff, especially in mountainous and ethnic minority areas, remain dependent on the support provided by higher level agencies.
- A new rural development model requires large resources, especially in terms of developing infrastructure. Particularly, the North-West region and Mekong River Delta, where infrastructure remains poor, require significant investments.
- Agricultural labourers make up 60 percent of the national workforce. This percentage must be reduced per the law of development. However, development of industry and service remains slow. Even trained rural labourers face difficulties in finding employment. Therefore, it will not be an easy task to lower the agricultural workforce to 30 percent by 2020, unless active and practical measures are defined and implemented.

- Raising the incomes and living conditions of rural inhabitants is essential and the ultimate objective of the national targeted programme. However, Vietnam faces a large gap between commodity production requirements and fragmented and small production land (average area of farming land is 1.6 ha per household nationally, 2.98 ha per household in the Northern Central Region, 2.13 ha per household in the Southern Central region, and nearly 0.35 ha per household in the Red River Delta, where intensive farming at a high level is applied, but the farming area of a household may be divided into 3 to 4 plots), and the gap between small farming and a large market. Therefore, it is necessary to restructure production, invest in a production infrastructure, train farmers, and apply new technological advances to agriculture. This, of course, will take some time to implement.
- Agriculture and rural areas are not as attractive to investors, as they pose many risks; regardless, there is the need to rapidly develop agricultural and rural enterprises in order to promote labourer distribution and increase high-value agro-products. At present, only 3.7 percent of enterprises work in agriculture, forestry, and fisheries sectors and the enterprises working in rural areas only invest 6 percent of their capital.
- Experiences in the new rural development model remain limited.

To solve these problems and overcome modern challenges, after two decades of implementing the *Doi moi* strategy, in 2008, the Vietnamese Communist Party decided that the new goal of rural economic development for the period 2011–2020 will be a development of new rural areas where modern economic, social, and service infrastructures are in place and ongoing improvement of the link between industries is focused.

The new rural development, hence, has five major characteristics:

- The communes will be civilised and clean, and infrastructure will be modernised.
- The production will follow market signals and be based on a market economy. Agriculture will be developed sustainably in order to protect the environment and ecological system in some forested areas.
- The physical and spiritual life of rural and agricultural households will be improved considerably.
- Ethnic cultural life will be maintained and developed.
- The rural society will be managed democratically and the security maintained.

In June 2010, the Vietnamese government approved the National Targeted Program for New Rural Development for the period 2011–2020. The programme is an effort on the part of government to implement Resolution 26 of the Vietnam Communist Party, which aims to help all communes reach 19 criteria.

As a result, some urgent activities have been selected by the Vietnamese government in order to properly implement the National Targeted Program for New Rural Development. These are as follows:

Firstly, to formulate rural planning and issue rural construction regulation in order to quickly deal with the ongoing unmethodical construction that consequently destroys the rural landscape and environment.

The new planning will include general planning (residential areas, socio-economic and production infrastructure) and detailed planning of the stated areas. Participating in the planning must include local communities, professional agencies, and district People's Committees, which will appraise and approve the planning so as to ensure modern rural development and cultural identity conservation (of 54 ethnic groups).

It is regulated by the government that public works will not be constructed unless commune planning has been finalised. The government will begin providing sufficient funding for finishing commune planning by the end of 2011.

Secondly, the economy is restructuring every commune under the effect of a market economy. In order to achieve this goal, professional agencies will provide communities with guidance for selecting the plants, animals, and industries that match the planning so that every commune will have some major commodities to improve their production effectiveness and income.

Thirdly, improving and developing the rural infrastructure. This is considered as a breakthrough for changing rural areas and promoting socio-economic development. The government has a policy to invest 100 percent in five essential infrastructure programmes, which include the main roads of communes, schools, clinics, cultural houses, and Commune People Committee headquarters (excluding two non-infrastructure programmes, namely training and planning staff). The government will provide partial funding for other infrastructure works (main roads of villages, water supply stations, commune stadiums, village cultural houses, village sports areas, rural electricity, markets, rural communication system, etc.).

Fourthly, training programme operators to expand knowledge training for farmers, so they can have the necessary knowledge (for agro-production, rural development, civilised life, etc.) for owning the new rural development.

Fifthly, prioritise investment in especially difficult areas (62 poor districts included in programme 30A), including difficult districts, communes, and villages not included in programme 30. This investment should also be prioritised for some good communes (which have realised the criteria) so that 20 percent of the communes can achieve the objectives and become models for neighbourhoods.

2.3.3. *The changing of the rural financial market*

The provision of agricultural and rural financial services has been a central tenet of poverty reduction and employment creation measures of Vietnamese government since *Doi Moi* in 1986.

The rural financial system is formed by three providers that can be classified as (i) formal (registered) credit institutions; (ii) semi-formal institutions (mainly NGO-MFIs); (iii) informal institutions.

Firstly, formal financial institutions are established by legal document such as the Law on Credit Institutions, Governmental Decrees, etc. They are for example banks, financial companies, leasing companies and People's Credit Funds. However, not all of them properly address the needs of poor people.

Secondly, semi-formal financial institutions in Vietnam are mainly funded by donors through aid for programmes or schemes managed by mass organisations, for example the Women's Union, the Farmers' Association, and the Labor Confederation.

Thirdly, informal financial institutions such as private money-lenders are formed by friends or relatives.

Rural financial services in Vietnam are dominated by banks despite a wide range of suppliers. This system is far removed from the needs of agricultural households, because farmers do not have many assets for using as a bank deposit to secure loans. Traditionally, banks serve their indoor clients and do not search for potential clients externally, especially not in the case of smaller clients. Under the new policy of the government, several banks have to provide loans for the poor. They are VBARD and VBSP.

The system can be depicted as follows:

Figure 2.3: Major agricultural and rural financial providers in Vietnam

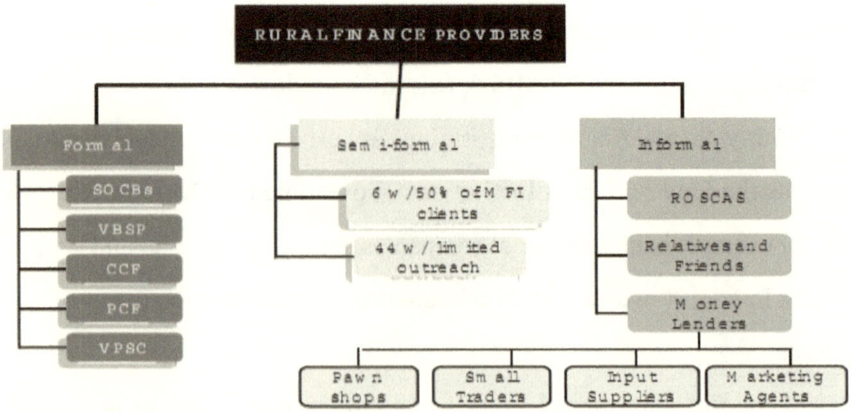

```
                    RURAL FINANCE PROVIDERS
        ┌──────────────────┬──────────────────┐
      Formal          Semi-formal          Informal
      ├─ SOCBs         ├─ 6 w/50% of MFI    ├─ ROSCAS
      ├─ VBSP          │    clients          ├─ Relatives and
      ├─ CCF           └─ 44 w/limited      │    Friends
      ├─ PCF                outreach         └─ Money
      └─ VPSC                                    Lenders
                    ┌────────┬────────┬────────┐
                   Pawn     Small    Input   Marketing
                   shops   Traders Suppliers  Agents
```

Source: ADB (2010)

According to the State Bank of Vietnam, formal financial institutions for rural development include six types:

- Commercial bank, especially VBARD. Prior to 2005, 16-plus commercial banks catered specifically to rural areas. After 2005, because the Vietnamese government does not distinguish between rural and urban banks, there were no more rural banks run specifically in these areas.
- Vietnam Bank for Social Policy (VBSP), 100 percent owned by the government, established in order to serve poor people with a fund source coming directly from the state budget.
- The People's Credit Funds (CDF) system with the Central People's Credit Fund (CCF) has been operated as a cooperative model.
- The Vietnam Postal Saving Company (CPSV) was established to supply mobilised saving services only. This company takes advantage of the appearance of its nearly 3,000 branches in all communes in Vietnam.
- TYM, a new type of NGO microfinance institution[6] registered in August 2010.

The operation of agricultural and rural financial suppliers can be presented as follows:

There are three large suppliers in rural financial markets: VBARD, VBSP, and CCF/PCFs. Among them, VBARD and CCF/PCFs are commercial financial institutions with funds coming mainly from saving

---

[6] For a detailed information on microfinance, please see Hung and Tam (2010).

on mobilisation at market rates; VBSP is a social bank that focuses on implementing government schemes or strategies for poverty alleviation with fund sources coming mainly from the state budget, of which the largest supplier is VBARD, with nearly 56 percent of the credit market. To date, VBARD has nearly 2,300 branches throughout the country; this bank concentrates on upper level customers of the rural financial market. The Vietnam State Bank has promoted PCFs to provide finance for farmers at the commune level, while CCFs have acted as apex institutions for PCFs and supporting PCFs in disbursing funds from the state bank. The purpose of the bank in maintaining the PCFs system is that the bank aims to restore the confidence in financial cooperatives that was damaged during the early period of *Doi moi* by hyper-inflation and the fast devaluation of Vietnamese currency.

By June 2010, there were more than 1,000 PCFs supporting about 1,000 communes (10 percent of total communes in Vietnam) and serving approximately 1.5 million members, 50 percent of which were poor. Despite the fact that PCFs are market-oriented, their operation is similar to the cooperative model. Fifteen percent of funding sources for PCFs comes mainly from CCFs. In 1995, the Vietnam Bank for the Poor (VBP) was established as a fund and managed by the VBA with the aim of serving poor customers. By 2002, the VBP had been organised independently, with VBARD becoming a non-profit institution called the Vietnam Social Policy Bank, which supplied financial support to poor farmers or disadvantaged groups as defined by the government. At the end of 2009, the VSPB had about 8,000 staff in total in all district branches, covering 98 percent of the communes in Vietnam (Hung & Tam, 2010).

This system is shown in the following figure:

Figure 2.4: Major institutions of the rural financial market in Vietnam

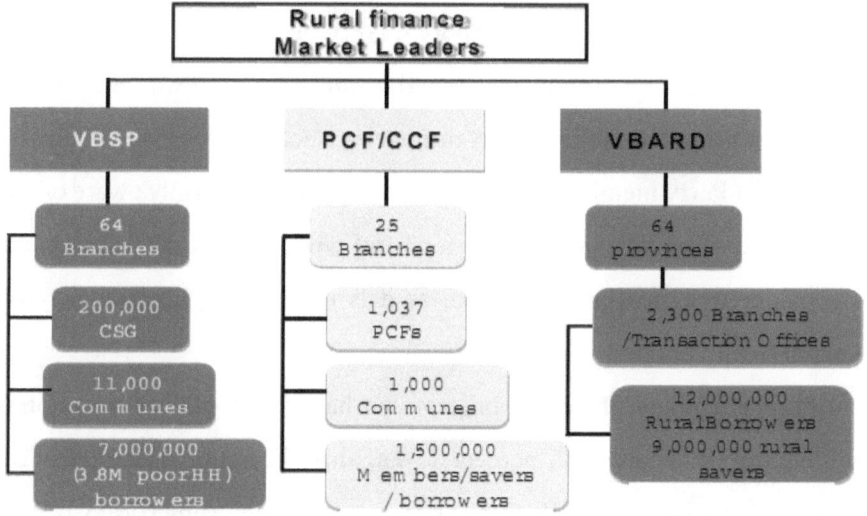

Source: ADB (2010)

2.3.4. The transformation of the labour market in the agricultural sector and rural areas and the role of vocational training

Following 20 years of reform, Vietnam's agriculture, farmers, and rural areas have achieved remarkable and tremendously comprehensive successes. However, these achievements are not yet commensurate with potential advantages and are uneven among regions. Economic restructuring and agricultural production renovation remain slow. Small and fragmented production, low productivity and quality, and low-added value of many products still remain.

Poorly-developed industrial and service sectors have not yet strongly promoted economic restructuring and rural labour force. Poor physical and spiritual life for rural people remains along with a high poverty rate, especially in ethnic minority and remote areas. The large social gap between rural and urban areas and among regions leads to social problems. One of the causes for this situation is the poor quality of the rural labour force.

For a long period of time, vocational training for rural workers has not been given proper attention. Many ministries, sectors, provinces, officials, Communist Party members, and communities are not yet fully aware of the importance of vocational training for rural workers. Vocational training is considered a short-term solution, as it is not regular, continuously or systematically applied.

The percentage of rural workers who have been offered vocational training thus far is only 18.7 percent, low against the national average of 25 percent. There is a large difference among regions regarding rural workers being offered vocational training courses (Red River Delta, 19.4 percent; Mekong River Delta, 17.9 percent; and the Northwest, only 8.3 percent). Vocational training policies targeted at rural labourers are still insufficient, deficient, and slow to be modified and amended. Vocational training networks are mainly developed in urban areas, whilst rural, mountainous, and remote areas have a very low number of vocational training institutions and small-scale, unqualified training conditions.

The state budget allocated to vocational training, in general and to rural workers in particular, has not yet met the requirements for expanding the scale and improving the quality of vocational training. The fund for the 'vocational training capacity building' project, under the national targeted programme for education by the year 2010, has been promptly increased over recent years; nonetheless, the number of vocational training centres and schools funded by the project remains low.

Many training institutions have been recently financed for 1–2 years. The project allocates funds for developing frameworks for intermediate and super-intermediate vocational training programmes. Funds have not yet been allocated to elaborate programmes, curricula, syllabi, and training materials for vocational elementary level and regular vocational level. The funds for rural worker training under the new project only support about 300,000 people per year. Support is low compared to actual requirements.

Aiming to achieve the party's strategic objective to develop Vietnam into an industrialised country by 2020, one of the important tasks is to develop and improve the quality of human resources, including for those who are offered vocational training. The Executive Committee for the Central Communist Party the Xth issued Resolution No. 26/NQ-TW, dated 5 August 2008, on agriculture, farmers, and rural areas (the so-called 'Resolution Tam Nong'). This resolution clearly indicates the party's vision and viewpoints on the comprehensive socio-economic development of Vietnam's rural areas under the national development strategy. One of the tasks specified in the resolution is to create jobs for farmers as a priority task throughout all national socio-economic development programmes, ensuring harmony among regions by narrowing the gap between regions and between rural and urban areas.

Implementing the resolutions through the Central Communist Party Committee, dated 28 October 2008, the government issued Resolution No. 24/2008/NQ-CP, promulgating the government's Action Plan, which focuses on training human resources in rural areas, shifting a part of agricultural labourers into industry and service sectors, creating jobs, and improving the rural population's income as high as 2.5 times that of what it currently is. One of the major tasks of the government's Action Plan is to develop the National Target Programme for training human resources in rural areas. This will concentrate on creating plans and solutions for training farmers' children, ensuring that they are qualified to work for industrial zones, handicraft and service sectors, and able to deal with job changes, should any occur. The remaining farmers who insist on agricultural production are to be trained in the necessary knowledge and skills to implement modern agricultural practices; focused training will also be provided to improve the knowledge of administrative officials and local officials.

In order to specify the Action Plan, based on the practical training programme as mentioned above, the Ministry of Labor, Invalids, and Social Affairs took the lead and coordinated with the Ministry of Agriculture and Rural Development, Ministry of Home Affairs, and relevant ministries, sectors, and mass organisations to formulate a project, 'Vocational training for rural workers by 2020'.

The Prime Minister issued Decision 1956/QD-TTg, dated 27 November 2009, approving the project 'Vocational training for rural workers by 2020' (hereafter called Project 1956). Under this decision, the party and she State consider vocational training for rural workers as the cause of the party and state, with different levels and social sectors aiming to improve the quality of rural workers, meeting requirements for industrialisation, agricultural modernisation, and rural areas. The state makes further investment in enhancing vocational training for rural workers by issuing policies for creating training opportunities for all rural workers in an attempt to ensure social justice and to encourage, mobilise, and facilitate the entire society to join the vocational training of rural workers. This is a legal framework for developing vocational training activities aimed at rural workers to improve the quality of rural human resources.

Project 1956 has set out training and advanced training for 100,000 commune officials. Aiming to meet the set objectives, the project has clarified measures and eight activities with a total state budget of 25,980 billion VND for 10 proposals. This has been the biggest project to be implemented for vocational training so far in terms of scale and budget. The project applies the mechanism to mobilise maximum resources and attract additional resources from international organisations, vocational training centres, enterprises, individuals, and communities.

Together with Project 1956, the government has issued Decision No. 800/QD-TTg, dated 4 June 2010, approving the national target programme on new rural development for the period 2010–2020. This is a master plan for socio-economic and political development and national security in rural areas. Accordingly, eleven issue groups will be implemented from the present to 2020, including further integration of industrialisation into rural development, job creation, and rural labour restructuring. As such, vocational training for rural workers thus far has been of greater concern to the state and the party than before. By then synchronous policies and solutions have been promulgated for the implementation.

Rural workers are the main labour force of Vietnamese society. However, at present, the majority of rural workers have not been properly trained. Vocational training for rural workers is the most efficient way for adopting science and technology in rural areas. Vocational training is the driving force for the success of new rural models and creates opportunities for making changes, restructuring the rural labour force and supplying human resources during industrialisation and modernisation. Vocational training for rural workers helps farmers create jobs, re-organise their production, and improve incomes and living standards in rural areas, which is considered an important criterion for new rural development.

Under the new rural development programme, the Party Commission Secretariat decided to select 11 pilot communes. When implementing the programme, the integration of goals is required. Resources are allocated to various programmes and projects, such as the national target programmes for poverty alleviation and the rural water supply programme. For the purpose of rural employment, Project 1956 will play an important role in three particular parts: (i) capacity building for farmers; (ii) organising vocational training for rural workers during non-harvest time (non-farm occupation); (iii) organise advanced training for local officials. As such, vocational training for the rural labour force will be the driving force for the success of new rural development.

For successful implementation, it is advised that the training methodology be changed, enabling farmers to select the training schools and subjects they expect to learn. Currently, the Ministry of Agriculture and Rural Development has issued vocational training certificates of farming practice as a pilot model. It is a way for creating opportunities for farmers to make their own decisions in the field of training. Moreover, vocational training for farmers is not required to be conducted by a state public school. Because if this, it is supposed to speed up socialisation and vocational training for farmers and to give priority to training those farmers who have had farming land withdrawn to be converted into industrial and urban land areas.

To effectively implement the project for vocational training for rural workers, citizens, non-governmental organisations, different local government bodies, mass organisations, and central government should work together on this strategy. In this way, this programme will be able to mobilise the participation of the entire political system from central to local levels.

Depending on socio-economic conditions, sectors and provinces can apply practical and creative activities to create many suitable vocational training models and approaches. Some models have initially attained achievements. Vocational training for rural workers not only mobilises 'brainstorming' in research institutes and universities, but also encourages technicians from enterprises and artisans from trade villages to be involved in teaching classes. Farmers and rural workers are policy beneficiaries, who have also proactively supported the government's policy by means of clarifying their training needs and fully participating in the training courses, since they find it useful and practical for themselves and their families. A number of pilot training models for labourers in some provinces (Lang Son, Cao Bang, Tay Ninh, Gia Lai, and others) have indicated that the improved skills of farmers have contributed to significantly improving labour productivity, crop quality, and income. The preliminary results have stimulated other rural employees to engage in training courses to be held at grassroots levels.

To make vocational training for rural workers a reality and create a consensus among local people, it should focus on the following issues:

The first, the vocational training programme for rural workers will be based on the employment demand of local enterprises and actual training needs of local people, rather than temporary and short-term activities.

Therefore, it is necessary to have a firm grasp of the training needs (categorised by occupation, professional groups, job positions, etc.) of local people in specific localities (commune, district) and of enterprises conditional to the training needs assessment surveys.

The second, vocational training needs the involvement of the entire political system. The reality shows that localities that are managed to receive strong support from the Vietnamese Communist Party executive committees and the sharp guidance of local authorities, and active participation of social and political organisations, will achieve the required results of vocational training for rural workers as expected.

The third, due to the cultural diversity of regions and the specific characteristics of farmers and rural workers (uneven education qualification, seasonal labour force, farming practices, etc.), training courses are expected to be organised in a flexible manner. These include training programs, training forms, training methods, communication methods, and locations (villages connected to fields of rural labourers), and attaching to such training time duration convenient to the farming practices of local people.

Finally, vocational training for rural workers will be linked to job creation, labour restructuring, poverty alleviation and social security in rural areas and new rural development.

To achieve this, close coordination is required between local government, training programme institutions and enterprises during the implementation process. The reality shows that good cooperation and coordination among relevant partners contributes to attaining positive achievements of vocational training, such as job creation and productivity improvement.

2.4. Policy debates and new directions for agricultural development in the future

At the time of writing, the debate concerning the agricultural development model in Vietnam continues. The debate focusses on the efficiency and equity implications of major institutional reform. The current debate goes back more than 50 years, when heated discussion existed between those who favoured a family farm model to develop rural economy and those who supported the 'Chinese model' of collectivised agriculture (the pro-Mao model). In the family farm model, production decisions are decentralised based on market price levels and follow the market signal; in the collectivised farming model, 'land is farmed by large brigades and run by cadres that assign the work, monitor progress, and allocate shares of net output to people according to the amount of work done' (M. Ravallion & D. Van De Walle, 2008, p. 24).

A large number of people supported the family farm model, but nonetheless lost the debate. The development of the collectivised farming model was partly a political ideology matter. Collectivisation pushed the agriculture of Vietnam into a classless situation where farmers had the same rights as their former landlords. In terms of practical tasks, collectivisation had helped to centre the financial resources for industrialisation at that time, as well as agricultural products for soldiers, their families and the war administrative system during the war between Vietnam and America. Additionally, proponents of collectivised farming argued that collectives could exploit the advantage of economic scale and reduce coordination problems. This is not a particularly convincing argument. For most crops, neither China nor Vietnam could apply capital-intensive technology in agricultural production through collectives, which is an indication of economics of scale. Another reason was that labour was abundant, meaning that agriculture in Vietnam did not face a scarcity of labour. Moreover, traditional villages had long since supplied goods and services to the market without forming collectives or cooperatives.

From 1988 to 1993, two versions of Land Law had been approved by the National Assembly of Vietnam: Land Law 1988 and Land Law 1993, respectively. This was followed by a series of discussions that are summarised by Kerkvliet and Selden:

> In Vietnam, the rights and obligations of rural landholders were spelled out in a 1993 land law passed by the national Assembly following extensive public debate. Significantly, not only Party officials but many villagers opposed privatization of land ownership rights. While favoring the long-term distribution of use rights to the fields, many preferred periodic redistribution in order to maintain equity, a pattern with roots in pre-revolution village praxis. (Benedict J. Tria Kerkvliet & Selden, 1998, p. 51)

There was a strong belief that collectives developed rural areas and provided opportunities for improving the living standard of farmers; in fact, collectives were deteriorating the agricultural sector in Vietnam. This belief remains in pro-Mao policy, but has been overcome by reformers who support moving toward a free market for agricultural land-use rights. Land Law 1993 created a new direction for rural development in Vietnam. This achievement is recognised not only by the government, but also by domestic and international scientists. It is argued that a dynamic land market has brought about more rapid poverty alleviation, though the policy allowed rich farmers to accumulate more land than inefficient farmers, facilitating diversification and increasing credit access for farmers (M. Ravallion & D. Van De Walle, 2008).

Under the industrialisation policy of the government, farmers have been attracted to the industrial sector; however, the policy has some side effects. Large areas of agricultural land have been transferred to industrial companies in order to build up new factories. These companies employ and train the farmers whose land they have bought. Initially, farmers benefited from this trend, because they received higher incomes. Higher incomes in turn absorb more agricultural labourers for the industrial sector. Consequently, in agriculture, only older people, women, and persons who were disqualified from working in industrial environments in the agricultural households are still harvesting. As a result, agricultural output does not reach maximum levels (Nguyen, 2004). Additionally, this situation has led to an easy accumulation of land for rich agricultural households and 'peasant classification' commonly occurs in the Vietnamese agricultural sector (Akram-Lodhi, 2004).

However, in times of financial crisis, especially during the international financial crisis of 2008, large numbers of labourers lost their jobs. When this happened, a significant number of labourers from factories and industrial companies flow back to the agricultural sector. Farmers sold their land to companies and became their employees; then, during the crisis period, farmers were made redundant and returned to their homes, but did not have any land. As a result, the problem of landlessness has been raised (Van de Walle & Cratty, 2004). In order for the economy to reach equilibrium during the crisis period, the government realised that they should maintain a sustainable agricultural sector to protect their people. Vietnam is able to escape an economic crisis easily with a stable agricultural sector.

Based on the experience gained in crisis times and pilot results presented in the above section, the government has decided to adjust the strategy of rural development. New rural development was approved in principle and has now been implemented throughout the country. The purpose of the new strategy is to fix problems caused by the industrialisation policy. As such, the government commits to several strategic activities:

The first, to raise the productivity of small farmers, the government should successfully implement farmer education programmes, essentially, vocational training programmes. Farmers need to be trained not only for their own benefit but also for the benefit of their potential employers in the industrial sector. Therefore, the government is spending money to build up trade schools and organisational industrial apprenticeships for farmers. All essential skills and knowledge concerning a market economy, working environment, business skills, and technical skills should be provided by the government. Other required skills will be supplied by the private sector, as these requirements depend on the demand of the companies.

Increasing the educational level of farmers will take time. Trainers need to be trained; courses must be conducted properly and reasonably; educational institutions must be developed. These processes may not be interesting to politicians, who like to have economic growth rates at 7–8 percent annually. Nonetheless, there are no shortcuts to achieving sustainable agriculture without preparation of farmers' skills and giving them essential knowledge about industrial works. In the past, the government has had bad experiences with compulsory collective farming.

The government now knows that forcing farmers into collective farming is not a wise policy for rural and agricultural development. This way of thinking is suitable with suggestions made by Lewis (1966).[7]

The second, to apply new knowledge or technology into agriculture, the government should let alone solutions that appear appropriate. Vietnam is a populous country and cannot ignore small farmers.[8] There are two essential reasons government should consider for this policy:

- If the government ignores small farmers and concentrate on large-scale farms, these will not run smoothly, because Vietnam still lacks labourers for establishing engineering service centres. Under the industrialisation policy, all skilled labourers and engineers are absorbed by industrial enterprises, leaving no mechanisms in rural areas that can provide engineering services for farmers. This is an administrative problem. The case of a shortage of skilled labourers in Vietnam is a difficult problem to solve.
- The other reason is the malfunction of science. Allowing for the concentration of resources will lead to scaling up virgin lands. Farmers have marginally explored virgin land over centuries. Each type of virgin land may have its own characteristics, such as erratic rainfall, difficult soils or lack of trace elements. Virgin land therefore poses a challenge to scientists who have studied

---

[7] Won a Nobel award for economics in 1979.

[8] In developed countries or under-populated countries, small farmers could be ignored, 'concentrating resources instead on opening up virgin lands in large-scale farms, using the latest machinery and the latest scientific knowledge' (Lewis, 1966, p. 36).

agricultural problems in cultivated lands. When suddenly sent to a virgin land with a crop will be cultivated immediately over thousands of hectares, they may not know what will happen if farmers apply their results of studying. The best solution in this case is for government to let farmers choose their own solution, because they have better knowledge of virgin land than others.

The third, in cases where 'farmers as actors of the development process' want to develop a new rural area, 'their patriotism, autonomy and self-reliance are aroused' to turn the new rural development programme into an emulation movement with the participation of the rural community in an attempt to (1) promote the pride and self-esteem of individual families and individual villages; (2) to involve all rural people in discussions and development vision for their own localities; (3) helping to change the operational mechanism of the programme; (4) to participate in monitoring the progress of constructional works; (5) to actively develop different forms of economic cooperation, cultural emulation, environmental protection, and social security. Therefore, the national targeted programme for new rural development is required to develop a new organisational structure and apply new policies. It is supposed to launch a campaign to disseminate the programme with a new approach.

Chapter 3

DATA DESCRIPTIONS

This chapter will present descriptions of datasets that are used in this thesis. In order to analyse the effects of investment on agricultural development, this thesis employed a mix of some various datasets that are released by the General Statistical Office in Vietnam. At the household level, Vietnam Household Living Standard Surveys in 2002, 2004, and 2006 have been employed to figure out the relationship between investment in health and education at the household level with their income. At the provincial level, to find out the effect of changes in public policy on productivity of agricultural firms, the thesis uses aggregated data for firms from 2005 to 2008. After that, firm dataset at the provincial level is matched with data on performance of public policy at the provincial level from 2005 to 2008, which is supported by the United States Agency for International Development (USAID) in collaboration with the Vietnam Chamber of Commerce and Industry (VCCI). This chapter, however, presents brief information about datasets, and detailed statistical information about variables will be presented in each chapter.

3.1. Vietnam Household Living Standard Survey

To evaluate living standards for policy-making and socio-economic development planning, the GSO conducts the VHLSS survey in every two years from 2002 to 2010. This survey is based on the two surveys which were done by the World Bank in Vietnam in 1993 and 1998. From 2002 to 2010, this survey was done in order to maintain the systematic monitoring and supervision of the living standards of different population groups in Vietnam. Another goal of the survey is to monitor and evaluate the implementation of the Comprehensive Poverty Reduction and Growth Strategy and to contribute to the evaluation of achievement of the Millennium Development Goals (MDGs) and Vietnam's socio-economic development goals.

3.1.1. Demography

The average number of household members in 2006 was 4.2 persons across the country (it was 4.44 persons in 2002 and 4.36 persons in 2004). This trend appeared in both rural and urban households, in all regions and different income levels. The average number of household members in rural areas and poor household is higher than urban areas and rich households.

3.1.2. Education

Based on the surveys, the literacy rate of population aged from 10 years old was 93.1 percent, which constituted a slight increase in comparison to previous years (92.1 percent in 2002 and 93 percent in 2004). The rate of people who have no qualification or have never gone to school from population aged from 15 years and up of the poorest quintile was 38 percent, 3.5 times higher than that of the richest quintile.

Average expenditure on education for one household member who went to school in the last 12 months in 2006 rose significantly in comparison with 2004. On average, households paid 1.211 million VND (equal to 75.5 USD) in 2006 for one household member, an increase of 47 percent in comparison with 2004. This was 32 percent distinction if we compare 2004 to 2002. School fees, expenditure on extra classes, and other educational expenditure accounted for a large fraction of educational expenditures for household members. Detail information is below:

Table 3.1. Average expense on education and training per person in the past 12 months by expense item, urban rural, region

Unit: 1,000 VND

			By expenditure item					
	Total	School fees	Contribution to school fund	Uniform	Textbook	Study tools	Extra class	Other expenses
1. Whole country								
2002	627	174	67	48	67	67	125	90
2004	826	253	86	60	89	67	130	99
2006	1211	347	75	68	105	85	180	230
2. Urban – Rural								
				2.1. Urban				
2002	1255	419	101	75	96	73	329	162
2004	1537	567	132	87	131	86	296	167
2006	2096	672	102	99	139	105	381	400

		By expenditure item						
	Total	School fees	Contribution to school fund	Uniform	Textbook	Study tools	Extra class	Other expenses
				2.2. Rural				
2002	433	99	57	40	57	50	62	68
2004	602	154	71	51	76	62	77	77
2006	894	230	65	57	93	78	107	169
3. Region								
				3.1. Red River Delta				
2002	711	221	70	23	75	58	171	92
2004	898	295	93	33	95	73	160	95
2006	1369	414	80	41	116	92	206	261
				3.2. North East				
2002	422	107	57	20	56	50	57	75
2004	616	187	93	23	74	62	70	86
2006	925	292	79	33	91	91	83	207
				3.3. North West				
2002	278	52	48	13	45	43	27	50
2004	294	69	55	11	37	45	14	47
2006	544	116	52	14	51	60	38	191
				3.4. North Central Coast				
2002	470	106	87	26	65	51	73	63
2004	650	156	116	32	86	62	94	70
2006	956	220	109	37	106	81	123	170
				3.5. South Central Coast				
2002	642	169	63	68	64	54	139	84
2004	807	235	75	89	87	68	125	115
2006	1236	360	61	97	110	87	162	228
				3.6. Central Highlands				
2002	513	103	96	56	56	48	84	71
2004	680	157	100	79	76	60	87	99
2006	958	208	92	84	104	85	102	170
				3.7. South East				
2002	1139	394	87	93	84	69	278	134
2004	1426	540	93	106	126	76	276	112
2006	2012	696	74	116	120	93	444	307

	Total	School fees	By expenditure item					
			Contribution to school fund	Uniform	Textbook	Study tools	Extra class	Other expenses
3.8. Mekong River Delta								
2002	491	99	33	79	59	56	62	105
2004	697	160	39	97	79	68	79	132
2006	934	207	35	106	96	82	97	225

Source: Author's calculation from VHLSS 2002, 2004, and 2006

On average, 2.096 million VND was paid for schooling of a person of urban households, 2.3 times higher than rural households. This ratio for 2004 and 2002 was 2.6 and 2.9 times, respectively. The largest part of educational expenses was tuition fees (it accounted for 32 and 26 percent of educational expenses in urban and rural areas, respectively, in 2006) and the second largest part was extra class (it accounted for 18 and 12 percent of expenditures on education in urban and rural areas, respectively, in 2006). The runner-up was expenses on textbooks. Voluntary contributions to school fund were right after and accounted for 5 and 7 percent of educational consumption in 2006, respectively). The last one was expenses on school uniforms. Additionally, extra classes were quite popular at the time of the survey, 2006. On average, about 43 percent of household members attended extra classes in the last 12 months, in which 68 percent of household members attended extra classes at schools and 28 percent of household members were at teacher's home. This rate of rich households and urban households was higher.

3.1.3. Labour, employment, and income

The number of economically active population trended positively in recent years. At the same time, active employment in young age groups has a trend to decline overtime. The percentage of active employment at 15–19 in total employment accounted for 12.6 percent in 2002, 11.3 percent in 2004, and 9.2 percent in 2006. It was a good indication because 15–19 population was school-aged. However, the percentage of this age group in poor households was higher than in rich households; urban areas were higher than rural areas. According to VHLSS 2002, 2004, and 2006, the percentages of the working group in economically active population of quintile 1 (the poorest households) were 16.5 percent, 15.6 percent, and 13.5 percent in 2002, 2004, and 2006, respectively. These percentages were far higher than the percentages of richest households (6.8 percent, 5.3 percent, and 3.8 percent in 2002, 2004, and 2006). Additionally, in 2002, the fraction of economically active population in the 15–19 age group in urban areas was only 6.9 percent in comparison with 14.3 percent in rural areas; in 2004 was 5.1 percent in comparison with 13.2 percent, and in 2006 was 4.5 percent in comparison with 10.9 percent, respectively. Detailed information is presented in the following table:

Table 3.2. Economically active population in working age by urban, rural, income quintile in 2002, 2004, and 2006

Unit: %

	15–19	20–24	25–29	30–34	35–39	40–44	45–49	50–54	55–59	60+
Age group										
1. Whole country										
2002	12.6	12.8	13.3	14.0	14.6	13.2	10.0	6.7	2.5	0.3
2004	11.3	11.3	11.6	13.1	14.2	14.5	12.1	8.1	2.3	…
2006	9.2	9.2	11.0	12.2	13.4	14.7	13.2	9.3	3.5	0.4
2. Urban – Rural										
2.1. Urban										
2002	6.9	11.3	13.9	14.5	15.3	15.9	12.1	7.4	2.5	0.3
2004	5.1	11.2	12.9	13.8	14.1	16.6	14.8	9.2	2.5	…
2006	4.5	11.8	12.6	12.6	13.4	15.3	15.5	10.4	3.6	0.3
2.1. Rural										
2002	14.3	13.3	13.1	13.9	14.4	12.4	9.4	6.5	2.4	0.3
2004	13.2	13.2	11.2	12.9	14.2	13.9	11.3	7.8	2.3	…
2006	10.9	13.7	10.5	12.0	13.4	14.4	12.4	8.9	3.4	0.4
3. Income quintile for the whole country										
3.1 Quintile 1										
2002	16.5	12.0	13.3	15.3	16.4	12.1	7.7	4.7	1.8	0.3
2004	15.6	11.9	11.1	15.3	16.3	13.9	8.6	5.8	1.6	…
2006	13.5	12.4	10.1	14.5	15.3	14.5	10.2	6.5	2.7	0.3
3.2. Quintile 2										
2002	15.3	12.2	12.8	14.4	15.6	13.0	8.8	5.7	2.1	0.3
2004	13.9	12.1	10.9	13.5	15.1	14.7	10.7	7.3	1.8	…
2006	12.0	12.6	9.4	12.7	15.4	14.8	12.0	8.0	2.8	0.3
3.3. Quintile 3										
2002	13.7	13.2	12.6	14.1	14.3	13.1	9.8	6.5	2.5	0.3
2004	12.5	12.9	11.1	12.6	13.8	14.5	12.0	8.2	2.4	…
2006	10.2	13.8	10.7	11.3	12.9	15.2	13.1	8.9	3.5	0.4
3.4. Quintile 4										
2002	11.5	14.1	12.8	13.0	13.7	13.2	11.0	7.7	2.7	0.4
2004	9.9	14.2	11.5	11.9	13.5	14.4	13.6	8.5	2.6	…
2006	7.4	14.5	11.8	11.2	12.2	14.2	13.9	10.6	3.7	0.4

	\multicolumn{10}{c}{Age group}
	15–19	20–24	25–29	30–34	35–39	40–44	45–49	50–54	55–59	60+
	\multicolumn{10}{c}{*3.5. Quintile 5*}
2002	6.8	12.5	14.8	13.6	13.4	14.3	12.5	8.6	3.1	0.3
2004	5.3	12.4	13.4	12.7	12.6	14.9	15.2	10.5	3.1	...
2006	3.8	12.6	12.8	11.4	11.6	14.5	16.2	12.0	4.5	0.4

Source: Author's calculation from VHLSS 2002, 2004, and 2006

In 2006, the monthly average income per capita at current price hit 636,000 VND, an increase of 31.4 percent compared to 2004. In the period 2004–2006, the annual increase of monthly income per capita at current price was 14.6 percent, lower than the annual increase of 16.6 percent for the period 2002–2004. If the increase in price was eliminated, the annual increase of real income per capita for the period 2004–2006 was 6.2 percent, lower than the annual increase of 10.7 percent for the period 2002–2004. The main reasons for income increase were the enlargement of production, plant's output increase, especially rice output; agricultural and fishery prices like rice, coffee, rubber, cashew, pork (live weight), shrimp, and fish increased in comparison with 2004 (GSO, 2008). However, the increase of income per capita for the period 2004–2006 was lower than the period 2002–2004 due to the lower increase of income from agricultural and non-farm production.

In addition, the structure of income had not changed much in the period 2002–2006. Out of total income, income from wages and salary accounted for 32.7, 32.7, and 34.3 percent in 2002, 2004, and 2006, respectively. Income from agriculture, fishery, and forestry was the second and accounted for 28.6, 27.2, and 24.8 percent in 2002, 2004, and 2006, correspondingly. Income from industry and construction accounted for 6.0, 5.8, and 6.0 percent in 2002, 2004, and 2006, in that order. The percentage of income from services in total income in 2002 was 16.7 percent, lower than the percentage in 2006 (16.8 percent) and remained the same percentage in 2004. Other income had increased moderately from 2002 to 2006, an increase of 1.9 percent from 2002 to 2006. Moreover, there was a different income structure between rural and urban households. The main source of income of urban households was salary or wage and services, while the major source of income of rural households was salary or wage and agriculture. This difference remained for the period of 2002–2006. Income from salary or wage of urban households accounted for 44.2, 42.5, and 42.9 percent in 2002, 2004, and 2006, respectively, while income from agriculture of rural households accounted for 36.0, 35.1, and 33.0 in 2002, 2004, and 2006, correspondingly. Detailed information is presented in the table below.

Table 3.3. Monthly income per capita by source of income and its structure by urban, rural in 2002, 2004, and 2006

	Total	Salary or wage	Agriculture	Forestry	Fishery	Industry	Construction	Trade	Service	Other
					Unit: 1,000 VND					
1. Whole country										
2002	356.1	116.4	82.4	4.5	14.5	19.7	1.6	34.4	24.8	57.7
2004	484.4	158.4	109.5	4.8	17.4	26.0	2.1	47.8	32.8	85.5
2006	636.5	218.0	132.0	5.1	21.2	34.8	3.5	62.3	44.5	115.1
2. Urban – Rural										
					2.1. Urban					
2002	622.1	274.7	28.0	1.0	13.7	35.8	4.8	76.9	66.4	120.7
2004	815.4	346.1	37.8	0.7	9.5	43.4	4.3	101.9	90.0	181.8
2006	1058.4	453.8	46.7	0.8	10.6	63.6	8.0	129.2	116.0	229.6
					2.2 Rural					
2002	275.1	68.2	99.0	5.6	14.8	14.8	0.6	21.5	12.2	38.5
2004	378.1	98.1	132.5	6.1	20.0	20.4	1.5	30.5	14.5	54.6
2006	505.7	140.0	167.1	6.8	25.8	25.6	2.0	40.3	20.6	77.6

	Total	Salary or wage	Agriculture	Forestry	Fishery	Industry	Construction	Trade	Service	Other
					Unit: %					
1. Whole country										
2002		32.7	23.2	1.3	4.1	5.5	0.5	9.7	7.0	16.2
2004		32.7	22.6	1.0	3.6	5.4	0.4	9.9	6.8	17.7
2006		34.3	20.7	0.8	3.3	5.5	0.5	9.8	7.0	18.1
2. Urban – Rural										
					2.1. Urban					
2002		44.2	4.5	0.2	2.2	5.8	0.8	12.4	10.7	19.4
2004		42.5	4.6	0.1	1.2	5.3	0.5	12.5	11.0	22.3
2006		42.9	4.4	0.1	1.0	6.0	0.8	12.2	11.0	21.7
					2.2 Rural					
2002		24.8	36.0	2.0	5.4	5.4	0.2	7.8	4.4	14.0
2004		26.0	35.1	1.6	5.3	5.4	0.4	8.1	3.8	14.4
2006		27.7	33.0	1.3	5.1	5.1	0.4	8.0	4.1	15.3

Source: Author's calculation from VHLSS 2002, 2004, and 2006

3.1.4. Health and health care

There was an increase in health expenditure per capita for both in-patient and out-patient treatment from 2002 to 2006. According to VHLSS results in 2002, 2004, and 2006, health expenditure per capita on in-patient treatment increased significantly from 1.43 million VND in 2002 to 1.65 million VND in 2004 and then 1.78 million VND in 2006. Health expenditure per person on out-patient treatment increased slightly from 0.35 million VND in 2002 to 0.36 million VND and then 0.38 million VND in 2006. Health expenditure per person on in-patient treatment in urban area increased considerably from 2002 to 2004 and then decreased slightly from 2004 to 2006, while the expenditure on out-patient treatment increased from 2002 to 2004 and then remained between 2004 and 2006. On the other hand, health expenditure per person in rural areas on in-patient and out-patient treatment increased over the period 2002–2006, especially for in-patient treatment. Detailed information is presented below:

Table 3.4. Average health expenditure per person having treatment in the past 12 months by type of treatment, urban and rural region

Unit: 1,000 VND

	2002		2004		2006	
	In-patient	Out-patient	In-patient	Out-patient	In-patient	Out-patient
1. Whole country						
	1430.2	352.7	1652.9	361.5	1787.9	377.8
2. Urban – Rural						
Urban	2055.4	528.3	2233.2	550.7	2187.6	550.6
Rural	1239.6	290.9	1467.1	291.7	1631.7	310.7
3. Region						
Red River Delta	1445.4	361.2	1639.4	334.9	2029.7	393.7
North East	1016.1	251.4	1040.7	226.9	1368.5	245.5
North West	745.9	190.9	1314.0	179.8	855.4	153.3
North Central Coast	1084.4	244.5	1114.2	317.3	1574.0	252.7
South Central Coast	1319.3	316.9	1496.8	310.4	1746.8	288.7
Central Highlands	1113.4	253.7	1294.4	257.2	1059.5	383.4
South East	2247.6	516.1	2959.3	581.3	2519.1	602.6
Mekong River Delta	1656.8	345.7	1781.5	341.1	1909.6	344.6

Source: Author's calculation from VHLSS 2002, 2004, and 2006

The common trend among regions is that health expenditure per person on in-patient treatment increased from 2002 to 2006 except for people who lived in Central Highlands, while the expenditure on out-patient treatment fluctuated across regions. The biggest change in health expenditure per capita on in-patient treatment was on people who were in Red River Delta, where the expenditure increased by 0.58 million VND from 2002 to 2006, while the lowest was Central Highlands people. However, the biggest change in health expenditure on out-patient treatment was people who live in Central Highlands. Health expenditures on out-patient treatment in North-East, North-West, and Mekong River Delta decreased considerably, while the other regions experienced an increase on the expenditure.

3.1.5. Characteristics of commune

In order to evaluate the effect of business production process on the improvement of living standards in rural areas, VHLSS 2002, 2004, and 2006 collected data from nearly 2,300 communes within the country. Villages and communes information included commune characteristics, chances for non-farm jobs, agricultural production situation of communes, infrastructure, education, health, credit, and saving.

According to the evaluation of key staff of communes, 99.1 percent of surveyed communes had a better living standard in comparison with five previous years. This rate was 98.7 percent in 2004 and 97.7 percent in 2002. The first cause of improved living standards of rural households was the increase of agricultural income (as indicated above). Productivity of most plants of communes increased because of the changing method of cultivating, expansion of watered agricultural land and more convenient conditions in exchanging agricultural products. Other causes of improved living standards were the increased income from non-farm businesses of households. In 2004 and 2006, there were 50 percent and 52 percent communes respectively with business production establishments, handicraft trade villages inside the commune or nearby communes. Moreover, the percentage of communes with business production, service establishments, and traditional craft villages inside the commune or nearby was 88 percent in 2004 and 89 percent in 2006.

Based on VHLSS survey results, most communes have a primary school (98 percent) and secondary school (90 percent). This percentage was highest in Red River Delta and lowest in the Central Highlands (respectively was 100 percent and 99 percent, in comparison with 90 percent and 83 percent). Additionally, means of travelling to school in communes were improved significantly. In 2004, 40 percent of primary schools had most pupils travel to school by bike and 56 percent travelled by foot. In 2006, these figures were 52 percent and 44 percent, in that order. One more point here is that the percentage of communes having at least one drop-out pupil decreased moderately from 2003 to 2005. This percentage in 2003 was 44 percent. In 2005, the percentage declined to 37 percent. The major reasons for dropping out were difficult economic conditions (60 percent), parents did not care about children's education (59 percent) and children could not learn or did not want to study (51 percent).

3.2. Vietnam Enterprise Survey

This study uses aggregated data for Vietnam's agricultural enterprises, drawn from the Vietnam Enterprise Survey (hereafter VES) conducted annually since 2001. These surveys were administered by the GSO using all enterprises employing 10 workers and above that were in operation at the end of the previous year. The list of enterprises was taken from tax departments and then the informal sector was excluded. The GSO's divisions at the provincial level were responsible for data collection from the listed enterprises by interviewing directly. Completed questionnaires were signed and stamped by enterprises' managers before submitting to GSO's divisions. The information reported by enterprises was assumed to be as precise as the official data that they reported annually to tax authorities.

VES provides a rich source of data at the enterprise level, including data on the enterprises' output, input (including number of employees), its age, its sector of operation and location, the style of ownership, whether enterprises or not the exporters, and whether or not they used information technology. The sample used for econometrics analysis in this thesis is confined to a sub-sample at the provincial level, which is focused on agricultural enterprises for a total of more than 500 observations. The general information on agricultural enterprises is presented in the table below:

Table 3.5. General information of agricultural enterprises from 2000 to 2008

	2000	2001	2002	2003	2004	2005	2006	2007	2008
1. Number of enterprises	42,288	51,680	62,908	72,012	91,755	112,950	131,318	155,771	205,689
2. Number of agricultural enterprises (AE)	3,378	3,438	3,379	2,407	2,369	2,429	2,399	2,807	8,619
3. Number of labour of AE (1000 workers)	230	263.4	265.8	252.1	256.1	259.1	256.4	253.3	407.2
4. Fixed capital and long-term investment of AE (billion VND)	19,689	22,483	26,492	27,589	30,659	33,841	35,713	38,374	50,320
5. Revenue of AE (billion VND)	10,389	9,517	11,750	13,210	17,225	20,586	26,147	30,184	40,116

Source: Author's estimation from enterprise surveys

For the purpose of this study, output is proxied by revenue and inputs are labour and capital, which is proxied by fixed capital and long term investment. Output is measured in billion VND, labour is measured in 1,000 workers, and capital is measured in billion VND. Besides labour and capital, some other variables proxied for weather pattern such as distance from central of province to parallel 17th and dummy variables.

3.3. Provincial Competitive Index Survey

The Provincial Competitiveness Index (hereafter PCI) is conducted by a collaboration between the Vietnam Chamber of Commerce and Industry (VCCI) and USAID, funded Vietnam Competitiveness Initiative (VNCI) Project, with a substantial contribution by VNCI partner The Asia Foundation (TAF). This project is led by Professor Edmund Malesky from the University of California at San Diego. Professor Malesky is in charge of constructing research methodology and the presentation of analytical findings.

To build up the sample, researchers obtain tax-paying firms in all provinces from the National Tax Authority. This list is preferred to the lists provided by the Department of Planning and Investment of 64 provinces. After that, VCCI branches in all provinces check the address, telephone, and other information which are provided by the National Tax Authority in order to have input information for selecting the survey sample. VNCI then mails out a questionnaire to enterprises in order to obtain information on the quality and performance of provincial authorities on economic governance from private sector businesses.

PCI has ten weighted sub-indices. The ten sub-indices are presented in the following table:

Moreover, PCI could be seen as a tool for measuring and quantifying the standards of economic governance in all provinces in Vietnam based on the voice of private sector businesses. Therefore, in this thesis, this dataset has been employed.

Chapter 4

HUMAN CAPITAL, SOCIAL CAPITAL, AND AGRICULTURAL DEVELOPMENT: A LITERATURE REVIEW

4.1. Introduction

This chapter provides a summary of literature on human and social capital and agricultural development. Human and social capitals have recently entered economic literature as two important sources of economic growth. Human capital has been accepted worldwide as an economic factor, while social capital has only just recently begun to be acknowledged. Particularly in agricultural development, the roles of human and social capital are now seen as major dynamics. In the literature, agricultural technology shocks such as the Green Revolution is observed as igniting long-term economic growth, while human and social capital maintain this growth process. Additionally, the capitals have played a significant part in the transformation of agriculture. Schooling, nutritional intake of agricultural labourers, and cooperation among farmers have been reported worldwide as main factors of agricultural development. Some studies focus on the effects of both human and social capital on agricultural development. In order to point out the role of investment in human and social capital, this chapter aims to link together literature on human capital, social capital, and agricultural development.

The role of human capital on economic growth is recognised worldwide, while the importance of social capital on the growth of economic productivity remains under scrutiny. The existing gap in the literature suggests that the determinants of human and social capital on agricultural development have as such not been comprehensively solved. The major purpose of this chapter is to try and explore the relationship between human capital, social capital, and agricultural development.

This chapter is organised as follows: (i) provide a summary of human capital; (ii) social capital is reviewed; (iii) an assessment of agricultural development literature is presented.

4.2. Human capital

4.2.1. Definitions of human capital

Definitions of human capital have been developed since the publication of a paper by Schultz in 1961. Schultz (1961) defined human capital as a combination of two forms of skills and knowledge of people.

Since then, a variety of definitions of human capital has been proposed. Becker (1962) identified human capital as a combination of on-the-job training, schooling, information and health. According to Becker, on-the-job training is the most important part of human capital: 'On-the-job training, therefore, is a process that raises future productivity and differs from school training in that an investment is made on the job rather than in an institution that specializes in teaching' (G. Becker, 1962, p. 11).

Côté and Healy (2001, p. 18) characterise human capital as 'the knowledge, skills, competencies and attributes embodied in individuals that facilitate the creation of personal, social and economic well-being'. Todaro (2009, p. 375) states, *'Human capital* is the term economists often use for education, health, and other human capacities that can raise productivity when increased', while Thirlwall (2006, p. 68) states that 'human capital refers to the skills and expertise embodied in the labor force through education and training'.

4.2.2. Methods of measuring human capital

There are three general methods of measuring human capital: cost-based, income-based, and education-based.

The first method, cost-based, is first introduced by Engel (1883) with the main idea regarding child rearing costs to parents. According to the researcher, the cost of child rearing is equal to the cost of raising a child until the age of 25; when the child reaches 26, they are fully produced human capital. His formula follows:

$$c_{xi} = c_{0i} + xc_{0i} + \sum_{1}^{x} k_i c_{0i}$$
$$= c_{0i}\left\{1 + x + \frac{k_i x(x+1)}{2}\right\}$$
(4.1)

where x is the age of the child and $x < 26$, i is defined as class with $i = 1,2,3$ for the lower, middle, and upper class, respectively; c_{0i} is the cost at birth and $c_{0i} + k_i c_{0i}$ is annual costs, with k_i being the discount rate. In his model, Engel assumes that $c_{01} = 100$, $c_{02} = 200$, $c_{03} = 300$, marks and $k_i = k = 0.1$.

Engel's method is criticised by Dagum and Slottje (2000). These authors suggest that Engel's method should not include estimation of individual human capital, because the method does not include the time value of money and the social costs invested in humans. To overcome this weakness, Engel's approach has recently been developed to a modern cost-based method for measuring human capital. The development incorporates the assumption that depreciated value of money spent on human capital defined as investment is equal to human capital stock.

Kendrick et al. (1976) and Eisner (1985, 1989) are two researchers who expand this method more systematically. Kendrick divides investment on human capital into tangible and intangible aspects, where tangible components are the cost of investing in a physical human being and intangible investments are the costs to improve the productivity of labourers. These expenditures involve health, education, mobility, training, and the opportunities cost of attending school. Unlike Kendrick, Eisner does not consider the cost of child rearing a component of human capital investment.

The second method, income-based, was first introduced by William Petty (1690). His method of calculating human capital was simple and did not account for population heterogeneity. He considered the human capital stock of England as the difference between estimated national income and property income. According to Kiker (1966), Farr (1853) developed the first strictly technical model of estimating the capital value of an individual. Farr tried to estimate an individual's human capital by calculating the present value of a human's future net income, adjusted for death, by using the life table. Based on Farr's idea, Dublin and Lotka (1930) formulate an individual's value at birth using the following equation:

$$V_0 = \sum_{x=0}^{\infty} \frac{S_{0,x}(W_x Y_x - C_x)}{(1+i)^x} \qquad (4.2)$$

Where i is the interest rate, $S_{0,x}$ is the probability that a person can live from birth to age x, W_x is the employment rate at age x, Y_x is the income of an individual from year x to year $x + 1$, and C_x is the cost of living.

The equation (4.2) expresses the idea of Farr, except it includes the assumption of unemployment rate rather than assuming full employment. This assumption is closer to real life than Farr's model. Moreover, the value of an individual at a particular age, for example, a, is that

$$V_a = \sum_{x=a}^{\infty} \frac{S_{a,x}(W_x Y_x - C_x)}{(1+i)^{x-a}} \quad (4.3)$$

Equation (4.3) can be expanded to the equation below:

$$V_a = \frac{(1+i)^a}{S_{0,a}} \sum_{0}^{\infty} \frac{S_{0,x}(W_x Y_x - C_x)}{(1+i)^x} + \sum_{x=0}^{a} \frac{S_{a,x}(C_x - W_x Y_x)}{(1+i)^{x-a}} \quad (4.4)^9$$

Correspondingly, the net cost of rearing an individual from birth to age *a* can be calculated as

$$C_a = \sum_{x=0}^{a} \frac{S_{a,x}(C_x - W_x Y_x)}{(1+i)^{x-a}} \quad (4.5)$$

Combining (4.4) with (4.2) and (4.3) yields

$$V_a = \frac{(1+i)^a}{S_{0,a}} V_0 + C_a \quad (4.6)$$

---

[9] This equation is formed from (4.3):

$$V_a = \sum_{a}^{\infty} \frac{S_{a,x}(W_x Y_x - C_x)}{(1+i)^{x-a}} = \sum_{x=0}^{\infty} \frac{S_{a,x}(W_x Y_x - C_x)}{(1+i)^{x-a}} - \sum_{x=0}^{a-1} \frac{S_{a,x}(W_x Y_x - C_x)}{(1+i)^{x-a}}$$

$$= \sum_{x=0}^{\infty} \frac{S_{0,x}(W_x Y_x - C_x)(1+i)^a}{S_{0,a}(1+i)^x} + \sum_{x=0}^{a-1} \frac{S_{a,x}(C_x - W_x Y_x)}{(1+x)^{x-a}}$$

$$= \frac{(1+i)^a}{S_{0,a}} \sum_{x=0}^{\infty} \frac{S_{0,x}(W_x Y_x - C_x)}{(1+i)^x}$$

Thus,

$$C_a = V_a - \frac{(1+i)^a}{S_{0,a}} V_0 \quad (4.7)$$

The equation (4.7) implies that the cost of producing an individual's human capital is the difference between his value at the age of a and the present value of his worth at birth. This difference is adjusted by probability of survival to age a. We can see that if we assume values at birth to be equal for everyone, then the cost of investment in human capital depends on the current value. The current value is different from person to person, because each individual has his own path of education, training, and health. Therefore, the high current value people have definitely high value of investment in human capital.

Jorgenson and Fraumeni (1989, 1992) present a comprehensive way of measuring human capital using an income-based approach. An important contribution of the two authors to this method is that they simplified the procedure of discounting future earnings to the present value. The value of a person at the age of a is equal to his/her current annual income plus the present value of his/her lifetime returns in the next period, adjusted by probability of survival. By employing the recursive process, Jorgenson and Fraumeni provide an innovative method with which value of lifetime income at the age of a can be calculated easily.

According to these authors, the lifetime income V of the gender s, at the age a in the year y with education attainment e can be calculated using the following equation:

$$V_{y,s,a,e} = Y_{y+1,s,a,e} + S_{y,s,a+1}V_{y,s,a+1,e}\frac{1+g}{1+i} \quad (4.8)$$

Y is annual income and S is the probability that a person can live another year, g is the growth rate of earnings and i is the interest rate. For the life cycle, Jorgenson and Fraumeni suggest five stages:

- Ages 0–4: No school and no work
- Age 5–13: School and no work
- Age 14–34: School and work
- Age 35–74: Work but no school
- Age 74+: No school or work

Incorporating the stages of the life cycle in the equation (4.8) is another contribution of Jorgenson and Fraumeni to the income-based method. People are involved in education only in the second and third stages of the life cycle, while the other three stages do not relate to schooling. Thus, equation (4.8) can be expanded with the new variable E as the school enrolment rate:

$$\begin{aligned}V_{y,s,a,e} = &Y_{y+1,s,a,e} \\ &+ \{E_{y+1,s,a,e}S_{y,s,a+1}V_{y,s,a+1,e+1} \\ &+ (1-E_{y+1,s,a,e})S_{y,s,a+1}V_{y,s,a+1,e}\}\frac{1+g}{1+i}\end{aligned} \quad (4.9)$$

Founded on the income-based approach, some researchers try to build up an index for human capital instead of monetary measure. Mulligan and Sala-i-Martin (1997) calculate human capital index in a specified state in a given year as the quotient of total labour income to uneducated workers' wage. Thus, the standard human capital h in the condition i at the time t is measured by

$$h_i(t) = \frac{\int_0^\infty w_i(t,s)\eta_i(t,s)ds}{w_i(t,0)} \quad (4.10)$$

where $w_i(t,s)$ is the wage of a worker with s years of schooling and $w_i(t,0)$ is the salary of an employee without education; $\eta_i(t,s)$ is the fraction of labourers with s years of schooling.

This method assumes that zero-schooling workers have the same human capital across time and space and do not necessarily have the same level of income. The rationale of this assumption is that workers with an education can have a different level of human capital due to the dissimilar quality and performance of education across regions. The quality and performance of education is expressed via workers' skills. For example, we cannot say a worker in the North of Australia has the same level of human capital as one in the South if they have the same education attainment, because the Northern worker may have better practical skills than the Southern employee.

Koman and Marin (1999) examined the cases of Austria and Germany by constructing an aggregate measure of human capital. For their method, human capital stock is weighted by the income of workers with respect to schooling levels. First, they calculate the number of individuals at age i and the highest level of schooling at time t is j as shown in the following equation:

$$H_{i,j,t} = H_{i-1,j,t-1}(1-\delta_{i,t}) + H^+_{i,j,t} - H^-_{i,j,t} \quad (4.11)$$

where $H^+_{i,j,t}$ is the quantity of people at age i who graduated education level j at time t, $H^-_{i,j,t}$ is the number of persons aged i who completed the highest level of schooling j at time $t-1$ and who graduated higher schooling at time t, and $\delta_{i,t}$ is the probability that people die before they reach age i from age $i-1$ at time $t-1$. Secondly, they employed the Cobb-Douglas aggregator to transmit workers to human capital h:

$$h = \ln\ln\left(\frac{H}{L}\right) = \sum_s \omega_s \ln\ln(\rho(s)) \quad (4.12)$$

where $\rho(s) = L(s)/L$ is the proportion of labourers with s years of schooling, $\omega_s = e^{\gamma s} L(s) / \sum_s e^{\gamma s} L(s)$ is the efficieny variable of an employee with s years of schooling and γs represents Mincerian wage regression.

The third method, the education-based approach, has been widely used in the research of Benhabib and Spiegel (1994), Barro and Sala-i-Martin (1995), Islam (1995), Barro (1997, 1999), Temple (1999), Wolff (2000), and Krueger and Lindahl (2001). Unlike the two above methods, the education-based approach calculates human capital through its outcomes. Literacy rates, enrolment rates, dropout rates, repetition rates, average years of schooling, and test scores are measured to determine the stocks of human capital at individual and aggregate levels.

Literacy rates have been used for controlling human capital in the growth equation in early research papers of Romer (1989) and Azariadis and Drazen (1990). However, this variable excludes many other aspects of human capital, such as numeracy and technological knowledge. Using literacy rates as the main proxy for human capital also does not take into account the role of skills and knowledge to output. Judson

(2002) reminds us that literacy rates can be used as an alternative to human capital in the case of a country where the population have little education, while it is not a good proxy for countries with a universal primary education.

School enrolment rates are captured as proxies for human capital in the growth equation in a number of studies, including Barro (1991), Mankiw et al. (1992), Levine and Renelt (1992), and Gemmell (1996). The main reason for using this variable as human capital is that school enrolment rates present the new stocks of human capital that will be added to the old stocks. Thus, enrolment rates can be seen as future human capital stocks of a country; moreover, these rates are easy to compare among countries worldwide.

Average years of schooling compute the accumulated educational investment of the present labour force. This way of measuring assumes that the human capitals of workers are balanced to the years of schooling they have achieved. There are three main methods to measure average years of schooling: (i) the census/survey-based estimation method; (ii) the projection method; (iii) the perpetual inventory method. The first method is used by Psacharopoulos and Arriagada (1986, 1992), where they use the following equation to estimate mean years of schooling at country level: $\bar{S} = \sum L_i D_i$, where L_i is the fraction of workers contestant with i^{th} level of schooling and D_i is the length of the schooling at i^{th} level of schooling. The second method is employed by Kyriacou (1991). He regresses the dataset provided by Psachoropoulos and Arriagada (1986) for 42 countries by using the following equation: $S_{1975} = \beta_1 + \beta_2 Prim_{1960} + \beta_3 Sec_{1970} + \beta_4 High_{1970}$, where *Prim*, *Sec*, and *High* stand for primary, secondary and higher education. Next, he uses the value of $\bar{\beta_i}(i = \overline{1,4})$ to predict the average years of schooling in the workforce for other years (1965, 1970, 1980, 1985) and other countries. The third method is introduced by Lau et al. (1991). These authors use the below equation to estimate the stock of education:

$$S_T = \sum_{T-a_{max}+6}^{T-a_{min}+6} \sum_{g=1}^{g_{max}} E_{g,t} \theta_{g,t} \quad (4.13)$$

where S is the total stock of education at the year T, $a_{min} = 15$ and $a_{max} = 64$, and $\theta_{g,t}$ is defined as the probability that a pupil in grade g at time t can live until the year T. Based on this idea, Nehru et al. (1995) adjust equation (4.13) to

$$S_T = \sum_{T-a_{max}+6}^{T-a_{min}+6} \sum_{g=1}^{g_{max}} E_{g,t} \left(1 - r_{g,t} - d_{g,t}\right) \theta_{g,t} \quad (4.14)$$

with $r_{g,t}$ the repetition rate and $d_{g,t}$ the dropout rate in grade g at time t. $r_{g,t}$ and $d_{g,t}$ are assumed constant over time, because these authors do not provide a good enough dataset.

Barro and Lee (1993) attempt to overcome the missing data problem by modifying the formula of Psachoropoulos and Arriagada (1986). They compute average years of schooling as

$$\overline{S} = D_p \left(\frac{1}{2} h_{ip} + h_{cp}\right) + \left(D_p + D_{s1}\right) h_{is} + \left(D_p + D_{s1} + D_{s2}\right) h_{cs}$$
$$+ \left(D_p + D_{s1} + D_{s2} + \frac{1}{2} D_h\right) h_{ih} + \left(D_p + D_{s1} + D_{s2} + D_h\right) h_{ch} \quad (4.14)$$

where h_j is the segment of the adult population with the highest education attainment j, $j = ip$ for incomplete primary, cp for complete primary, is for incomplete secondary, cs for complete secondary, ih for incomplete higher education, and ch for complete higher education. D represents duration of the ith level of schooling, i is valued as p, $s1$, $s2$, and h for primary, first cycle of secondary, second cycle of secondary, and higher education, respectively.

Quality of schooling has been placed in the growth equation as an indicator for variations in performance of human capital. Barro and Lee (1996, 2001) can be seen as offering the first attempt to explain the relationship between quality of schooling and economic growth. These authors calculate quality of schooling through: (i) public educational spending per student; (ii) pupil-teacher ratios; (iii) salaries of teachers and the duration of schooling; (iv) repetition and dropout rates; (v) international test scores of high school students and adults. However, too many quality measurements of schooling have brought about confusion about understanding the correlation between quality of schooling and economic growth. To overcome this problem, Hanushek and Kimko (2000) built up a unique measurement of quality of schooling by combining all available information of international mathematics and science test scores. Wößmann (2003) improves the estimation of human capital stock by incorporating Hanushek and Kimko's (2000) measurement into the following equation: $h_i^Q = e^{\Sigma_a r_a Q_i S_{ai}}$, where r_a is the world rate of return to education at the level a, 20%, 13.5%, and 10.7% for primary, secondary, and higher levels (from Psacharopoulos, 1994), Q_i is QL2 in the measurement of Hanushek and Kimko (2000), and S_{ai} is the average years of schooling for a population aged 15 and above (from Barro and Lee, 2001).

4.2.3. Empirical works

Based on the three above methods, many empirical studied have been conducted to calculate human capital stock.

Firstly, relying on a cost-based approach, Schultz (1961) estimated the educational stock of the labour force for the United States of America (US). Schultz found that the stock increased about 8.5 times from 1900 to 1956. Following on, Kendrick (1976) and Eisner (1985, 1989) enhanced the way of calculating human capital stock. Kendrick estimated US wealth and discovered that human capital was the main source of economic growth in the US from 1929 to 1969. According to the author, human capital stock always exceeded non-human capital stock in this period. He also finds that in 1969, the US's non-human capital stock was $3,200 billion, while human capital stock totalled $3,700 billion. In the period 1929–1969, the average rate of growth of human capital was 6.3 percent per year, whereas non-human capital stock expanded by only 4.9 percent annually. Educational capital stock accounted for about 40–60 percent of human capital and this segment increased consistently over time. Eisner (1985) expanded the Kendrick method to calculate the US's human capital stock in the period 1945–1981. Eisner found that in 1981, nearly 45% of total capital was human capital. Moreover, from 1945 to 1981, human capital in the US increased 4.4 percent per year, while non-human capital rose at a slower rate, 3.9 percent annually. In short, Kendrick's (1976) and Eisner's (1985) results were quite similar, except Kendrick's calculation of human capital was often higher than non-human capital, while Eisner's estimation was the reverse.

Secondly, the first attempt to calculate human capital stock using the income-based method was effected by Petty (1690). He estimated England's total human capital at £520 million, or £80 per capita. Farr (1853) valued £150 as the average human capital net of a farmer. Nicholson (1891) applied this approach to value the stock of human capital for the United Kingdom (UK) and found that living capital stock was five times higher than the stock of conventional capital. Moreover, in Australia, Wickens (1924) calculated the value of the services males and females bring to society. According to the author, males and females contributed £133 and £65 to society, respectively. He also found that in 1915, Australia had £6,211 million of human capital or £1,246 per capita. In the US, Jorgenson and Fraumeni (1989) estimated that the USs stock of human capital increased from $92 trillion in 1949 to $171 trillion in 1984, in 1982 constant dollars. The two researchers showed that human capital was 12 to 16 times higher than physical capital with respect to size. In addition, Jorgenson and Fraumeni's (1992) estimation of human capital was 17.5 to 18.8 times higher than Kendrick's.

Other researchers have adopted Jorgenson and Fraumeni's method to calculate the stock of human capital in several countries. Wei (2004) applied this method for Australia with minor adjustments. He focused only on two life cycle stages, from 25 to 65 (work and study/work only), and used four levels of education rather than the 18 levels of Jorgenson and Fraumeni's framework. Wei proved that human capital and educational attainment have a positive relationship in the context of Australia. According to him, the stock of human capital of a population aged 25–65 increased from $3.2 trillion in 1981 to $5.6 trillion in 2001, in 2001 prices. Wei believes that the increase in human capital was due to the increase in educated individuals. In his study, women accounted for approximately 40% of the total stock of human capital. Ahlroth et al. (2005) applied Jorgenson and Fraumeni's model to Swedish data and found that this method worked particularly well with micro data sets, such as a household living standard survey. They found that in the context of Sweden, human capital accounts for a six to eight times higher rate than physical capital.

Based on Jorgenson and Fraumeni's lifetime approach, Haizheng Li et al. (2009) estimated the human capital stock of China from 1985 to 2007. The authors found that the total Chinese human capital increased from nearly ¥27 billion in 1985 to approximately ¥118 billion in 2007, meaning that the average rate of growth of Chinese human capital is 6.78% per year. Dividing this time period into two sub-periods, 1985–1994 and 1995–2007, researchers discovered that the rate of growth of total human capital stock increased from 5.11% in the first period to 7.86% in the second period. Indeed, human capital per capita in China increases at an average annual rate of growth of 6.25%. For the first period, this rate was 3.9%, while in the second period the rate was 7.5%. Therefore, they concluded that after 1995, human capital was the main driving force for economic growth in China.

Thirdly, Barro, and Lee (1993) provided a method for filling in the missing data at the country level. They built up a panel data set for 129 countries over a five-year period from 1960 to 1985. In their study, South Asia rated the lowest region in terms of years of schooling and had the highest inequality with respect to education. The ratio of female to male students increased from 0.28:1 in 1960 to 0.48:1 in 1985, while in OECD countries this ratio was substantial at 0.94:1 for the same period. In a research paper by Tran and Do (2007), the researchers used the data of 64 provinces in Vietnam from 2000 to 2004. They found that at the provincial level, educational attainment had a strong and positive effect on economic growth. According to their research, a 1% increase in years of schooling will raise GDP levels by 0.16%. In addition, Tran and Do concluded that in Vietnam, the effect of human capital on economic growth is different among different regions. For example, 'human capital is very positively important in the South East and the South Central Coast, but it has negative impact on [the] GDP level of the Red River Delta.'

4.2.4. Discussion

Despite the fact that the above discussed three methods are very common, they nonetheless pose several limitations. This part of the chapter will provide a critical review of the three methods.

At the outset, the cost-based approach has four main drawbacks. Firstly, the value of capital in this method originates from its demand instead of from cost production. When we use this method, human capital may be overestimated. For example, the less healthy child will be raised at a higher cost than the healthy child. Secondly, there is a problem in classifying types of expenditure that should be considered either investments or consumptions. For instance, Kendrick assumes that all costs of breeding children to the age of 14 can be seen as investment, while Bowman (1969) argued that such types of expenditures should be seen as investments only in the context of slaves. Machlup (1984) asserts that basic expenditure should be regarded as consumption rather than investment. There is consensus among economists that the line between investment and consumption with respect to 'expenditure on man' is very small; therefore, the term 'expenditure on man' can be considered a controversial economic term.

The next limitation is the depreciation rate of human capital stocks. Normally, researchers apply simple tax accounting principles to this issue and consider that human capital depreciates in a similar manner as physical capital. Hence, Kendrick applied the adjusted double declining balance method, whereas Eisner applied the straight-line method. Their results did not calculate the appreciation of human capital at younger ages, or depreciation later in the life cycle as noted in the discussion of Mincer (1958, 1974). Graham and Webb (1979) found strong evidence for appreciation of human capital when they measured human capital stocks in the US. Finally, Jorgenson and Fraumeni (1989) point out that the cost-based approach does not consider the value of non-market activities.

Secondly, the income-based approach relies heavily on the assumption that differences in wages accurately reflect differences in productivity; therefore, if this assumption does not match, the model will fail. In fact, differences in wages may be caused by factors other than productivity, such as educational attainment or health capacity (for a more detailed discussion, see Strauss and Duncan, 1998). Other evidence for this drawback is that wage rate can be raised because of trade unions, or the rate might fall during a downturn period. In these situations, this method will supply a bias and nonsensical results. Another disadvantage is maintenance cost. Some authors dispute that human capital should be considered similar to physical capital, and that human capital should instead have net maintenance cost. De Foville (1905) and Eisner (1988) analysed this method and concluded that not deducting maintenance cost from gross earning can lead to overestimating human capital. Weisbrod (1961) attempted to calculate maintenance cost but faced many difficulties doing so. For example, the particular types of expenditure that should be classified as maintenance costs are not easily monitored.

The next shortcoming of this method is the availability of data on income. In the case of developing countries, this problem is more severe. In these countries, wage rates are not openly observable; unreliable data should be considered carefully when applying this method to developing countries. Indeed, Jorgenson and Fraumeni's model tries to overcome these shortcomings, but their method has attracted some criticism. Rothschild (1992) negates their important assumption that human capital raises the productivity not only in the workplace but also during leisure time. Dagum and Slottje (2000) point out that Jorgenson and Fraumeni's model has a bias problem because it does not consider the differences from person to person due to the differences of nature and nurture among individuals of the same sex and educational attainment. This method forgets informal schooling and assumes that all types of formal education with the same length should have equal rates of return. These problems of Jorgenson and Fraumeni's model cause a biased estimate of future income and, as a result, human capital.

Finally, although education-based measures are convenient in many ways, this method is criticised for not adequately representing human capital. Initially, education-based methods focus too much on quantity and do not maintain a balanced view between quantity and quality of education. This drawback has led to biased results in interpreting human capital with respect to education. In addition, many researchers point out that this method is inconvenient for comparisons at the macro and micro levels, their findings indicating that education affects the development process differently at each stage of development. Therefore, using education as a main proxy of human capital may lead to contradictory conclusions. In trying to measure quality of schooling, Barro and Lee (1996, 2001) provide a comprehensive method; however, their work is constrained by the fact that quality of schooling is a multifaceted variable. The two economists try to employ as many quality indicators as they can into their framework, but the results failed to satisfy their expectations, because estimating across indicators supplied very poor correlation.

4.3. Social capital

4.3.1. Definitions of social capital

There are many definitions related to the term 'social capital'. These have led to confusion about what constitutes 'social capital'. In the most general sense, social capital can be understood as presenting situations where individuals can use their rights within a group or networks to secure or enhance their benefits. Pierre Bourdieu (1986) states

> Social capital is an attribute of an individual in a social context. One can acquire social capital through purposeful actions and can transform social capital into conventional economic gains. The ability to do so, however, depends on the nature of the social obligations, connections, and networks available to you. (Cited from Sobel, 2002, p. 139)

Around this idea, Glaeser, Laibson, and Sacerdote (2002) built an economic model of social capital. The researchers define 'individual social capital as a person's social characteristics—including social skills, charisma, and the size of his Rolodex—which enable him to reap market and non-market returns from interactions with others' (p. 438). However, the definitions of social capital have been exacerbated by the word games of researchers. Social capital, hence, relates to social energy, community spirit, social bonds, civic virtue, community networks, social zone, extended friendships, community life, social resources, informal and formal networks, good neighbourliness, and social glue.

Having a consensus regarding a definition for social capital would help in its calculation and application within a public policy context. However, the multifaceted nature of the term makes it hard to summarise in a single sentence. Putnam (2000) argues that social capital has 'forceful, even quantifiable effects on many different aspects of our lives' (p. 27). These quantifiable effects include better health (Wilkinson, 1996), better educational achievement (Coleman, 1988), less corrupt, and more effective government (Putnam, 1995) and boosting economic growth (Fukuyama, 1995).

Portes (1998) believes that social capital is not a new sociological term. Bourdieu (1986) identifies a means of valuing social capital and Coleman (1988, 1990) offers a comprehensive theoretical framework on social capital. Bourdieu focused on the different effects of the forms of capital to unequal power of citizen classes, while Coleman (1990) used rational behaviours as a starting point and noted. For Coleman, 'social capital is defined by its function; it is not a single entity, but a variety of different entities having characteristics in common: they all consist of some aspect of a social structure, and they facilitate certain actions of individuals who are within the structure' (p. 302). Coleman (1988) points out that social capital has three forms. Firstly, obligations and expectations rely on the trustworthiness of the social environment and the actual extent of obligations. Secondly, information can flow properly through the channels of information inside social structures in order to provide a basis for action. Finally, the presence of norms went with effective sanctions. On the other hand, Fukuyama (1995) provides an economic framework where social capital and trust are treated as economic factors rather than social or political aspects, as is the case with Coleman or Putnam's works. Passey (2000) advocates that Fukuyama's aim is to try to explain the difference among countries in terms of economic performance originating from the basic differing levels of trust. According to Passey, the past level of trust in a certain country significantly affects its prosperity and degree of democracy, as well as its ability to compete economically.

Social capital can be understood as the shared knowledge, understandings, norms, rules, and expectations driving the actions of groups of individuals to achieve their interests (Coleman, 1988; Ostrom, 1990, 1992; Putnam, Leonardi, and Nanetti, 1993). According to these authors, participants in the networks are more productive with whatever physical and human capital they draw on if they have an agreement in place that they will coordinate activities and involve one another closely and trustfully to the sequence of future behaviours. Based on this definition, international organisations such as the World Bank and the Organization for Economic Co-operation and Development (OECD) have carried out their definitions on social capital. Côté and Healy (2001) define social capital as 'networks together with shared norms, values and understandings that facilitate co-operation within or among groups'. In this definition, networks relate closely to the objective of actors who are involved in networks. Members of networks share their own views, values, and knowledge in order to build up common norms, values, and understandings. The largest networks are therefore international agreements or organisations and smaller networks are governments. Cultural context is very important background information for understanding the choices of individuals and groups under the driving force of shared norms, values, and understandings. When members of networks have dissimilar understandings, they usually do not have agreement.

Social capital takes many different forms. Putnam, Leonardi, and Nanetti (1993) define networks, norms, and social beliefs as three forms of social capital, while Coleman (1988) identifies social capital as involving the three types mentioned above. Woolcock (1999) states that social capital has three basic forms: social bonds, bridges, and linkages. According to him, bonding refers to family relationships or ethnic groups. Bridging relates to the relationship between friends, fellows, and colleagues. Linking expresses the connections among social strata in order to share power, social status, and wealth by different groups.

In addition, the core of the social capital concept is trust. Some researchers see trust as an outcome of social capital (Woolcock, 2001), while others believe trust to be an element of shared values; others consider that trust can be seen as both (Côté and Healy, 2001). Dasgupta (1999) defines trust 'in the context of an individual forming expectations about actions of others that have a bearing on this individual's choice of action, when that action must be chosen before he or she can observe the actions of those others' (p. 330). Therefore, in the networks, individuals usually compare what they know about others and what others talk about and do. Members of networks cannot have trust if they know that members say one thing, but do something else. For example, citizens in particular areas will not trust local policemen if they know local policemen may be involved in corruption. Côté and Healy (2001) suggest that trust should be distinguished in three types: (i) 'inter-personal trust among familiars'; (ii) 'inter-personal trust among "strangers"'; (iii) 'trust in public and private institutions' (p. 41). According to the two authors, trust should be treated as a time lag variable.

The concept of social capital can be seen as the combination of two activities: 'social and civic participation' and 'networks of co-operation and solidarity'. Therefore, social capital can be considered a multifaceted concept, with social cohesion, trust, reciprocity, and institutional effectiveness being its main dimensions. Woolcock (2001) points out that social capital has at least seven components: families and youth, schools and education, community life, work and organisation, democracy and governance, problems related to collective action, and economic development. Recently, social capital has been upgraded with further elements that include physical and mental health, immigration, and public protection. Moreover, social capital has been developed to the concept of social ties.

In research by Glaeser et al. (2002), social capital is defined as 'a person's social characteristics—including social skills, charisma, and the size of his Rolodex[10]—which enables him to reap market and non-market returns from interaction with others' (p. 438). According to these authors,

---

[10] This is a trademark and refers to a small desktop file containing cards for names, addresses, and phone numbers.

individual social capital should be seen as a component of personal human capital. Bowles and Gintis (1998) argue that social skills belong to personal human capital and that schooling improves these skills. Hence, there should be common ground between social and human capital.

Ostrom (1999) points out,

> To create social capital in a self-conscious manner, individuals must spend time and energy working with one another to craft institutions—that is, sets of rules that will be used to allocate the benefits derived from an organized activity and to assign responsibility for paying cost. (p. 178)

Ostrom's definition of social capital has led to some ideas about the utility and expenditure functions of individuals when they decide to create a network or group or organisation. Moreover, the process of building a network is not done by itself. Instead, potential members of a particular network initially have to work hard to establish the rules of the network. The sets of rules have to cover all benefits, rights, and responsibilities of current and future members of a social network or group or organisation. Additionally, it is understood that if members have done something that contradicts the rules, they will be punished by other members of the network. For example, members of the WTO will be punished if they violate the agreement of the WTO. Another point is that members of social networks should consider carefully all aspects involved in joining a network prior to doing so, as their decision will take time and other resources to maintain in order to remain on track with the rules of the network.

4.3.2. Methods of measuring social capital

Measuring social capital is difficult. The core elements of social capital are trust and levels of engagement. Putnam, in his celebrated book on social capital, *Bowling Alone: The Collapse and Revival of American Community* (2000), employs a wide range of datasets covering both cross-sectional and longitudinal types. His composition index has five elements: (i) intensity of involvement in community and organisational life; (ii) public engagement; (iii) community and volunteering; (iv) informal sociability; (v) reported levels of inter-personal trust (according to Côté and Healy, 2001). Coleman (1988) developed a system of social capital indicators for children's education achievement covering personal, family, and community dimensions. According to him, social capital is measured in the context of creating human capital and should be categorised into two types: social capital in the family and social capital outside the family. He argues that every family has three types of capital: financial, human, and social capital. Financial capital relates to the wealth or income of the family, human capital refers to the educational levels of the parents and social capital in the family is the relationship between children and parents. Coleman put children at the centre of measurement of social capital of family, because childhood has an important role in creating human capital as it relates to children growing up. He points out that this progress depends on two main factors: the physical presence of adults and the attention paid to children by adults.

Measuring social capital under the context of public policy, the Canadian government's project on social capital has classified this type of capital into three categories: micro-approach, meso-approach, and macro-approach. Firstly, at the micro level, social capital is referred to as the value of collective action and this way of measuring employs game theory as a framework. Secondly, at the meso level, social capital is seen as instrumental value that has the role of allocating resources such as information and support services to the members of certain networks. Burt (1984), Lin (2001), and Portes (1998) address social capital at this level and agree that the structures of social networks can improve cooperation. Moreover, Grootaert and van Bastelaer (2001) use the term 'structural social capital' for the meso-level in a report of the World Bank on social capital. Finally, at the macro level, social capital has been closely related with the value of integration and social cohesion. This approach emphasises the role of trust and reciprocity.

In a report of the Canadian government, carried out by the Policy Research Initiative, two ways of measuring social capital are addressed: (1) measuring social capital through network structure; (2) measuring social capital through network dynamics. According to the report, the method measuring social capital should focus on the 'utility of specific resources' and 'their potential accessibility'. Another development of scholars that are similar with the approach suggesting by the report, Côté and Healy (2001) established two principles for measuring social capital. They write,

> In principle, measures of social capital should be *i)* as comprehensive as possible in their coverage of key dimensions (networks, values and norms); and *ii)* balanced

> between attitudinal or subjective elements on the one hand (*e.g.,* reported levels of trust) and behavioural aspects on the other (*e.g.,* membership of associations and extent of social ties). (Côté & Healy, 2001, p. 43)

Instead of assuming a common value that can be accessed by all members of the network, Franke asserts that 'measuring the social capital of an individual or a group does not mean attributing a value to all the resources that are useful in a particular situation and that can be mobilized at a given time' (S Franke, 2005, p. 111). In fact, at a given time, some members of a network needing a particular resource does not mean that all members of the network require this resource. Hence, the value of social capital does not increase if these members possess that resource. For example, poor members of society need assistance from their government, while other members may not need this type of social support.

Moreover, the two researchers advocate that applied scientists should consider the cultural context of individuals in empirical exercises. It is said that family background can be seen as a good proxy for this consideration, which is similar to Coleman's (1988) proposal.

4.3.3. Empirical works

There have been many works on measuring the relationship between social capital and factors such as performance of schooling, health, mortality, crime, institutional efficiency, and economic development and growth. I have chosen to focus on economic development and growth.

Among prominent empirical authors, Knack and Keefer's 1997 paper is seen as one of the most well-known empirical studies in measuring the relationship between social capital and economy. The two authors built up a dataset using 29 countries and comparing them to answer three research issues: (i) the correlation of trust, civic cooperation, and economic performance; (ii) the conflict between Putnam's (1993) and Olson's (1982) hypotheses; (iii) the determinants of trust and norms of civic cooperation. According to the authors, their paper delivers three important findings:

> First, trust and civic cooperation are associated with stronger economic performance. Second, associational activity is not correlated with economic performance-contrary to Putnam's [1993] findings across Italian regions. Third, we find that trust and norms of civic cooperation are stronger in countries with formal institutions that effectively protect property and contract rights, and in countries that are less polarized along lines of class or ethnicity. (Knack & Keefer, 1997, p. 1251)

Knack and Keefer use growth of income per capita and investment rate as two major proxies for economic performance variables and employ three variables: built up social capital called trust, norms of civic cooperation, and the mean number of memberships per capita from the World Values Survey[11] in 1981 and 1991. Trust is measured as in Paxton's research, but Knack and Keefer acknowledge that the meanings of people's

---

[11] The dataset can be downloaded from http://www.worldvaluessurvey.org.

survey answers are different. The mean of trust variable is 35.8% and the standard of error is 14%. The second variable, norms of civic cooperation, is measured by score (the maximum score is 50 points). This score is built up from five order answers.[12] The mean score is 39.4 and the standard of error is only 2. These two variables are used to answer the first issue. It is found that trust and norms of civic cooperation have a positive relationship with economic performance. Additionally, to test the second issue, Knack and Keefer (1997) break down the third variable into two groups called P-group[13] (present for Putnam hypothesis) and O-group[14] (present for Olson hypothesis). The empirical result indicates a situation where the O-group does not have any significant effect on the growth or investment equations, while '"Putnam" groups appear to harm investment' (Knack and Keefer, 1997, p. 1274). The result does not vary much when Knack and Keefer leave religious and church organisations out of the P-group and political parties or groups from the O-group. To explain this result, the two authors state that the positive effect of the P-group may be 'counterbalanced' by the negative effect of the O-group.

---

[12] Each question has three alternatives: always/never/sometimes is justified. Questions are '(1) claiming benefits you're not entitled to; (2) avoiding a fare on public transport; (3) cheating on taxes; (4) keeping money found; and (5) failing to report accidental damage cause to a parked vehicle' Ponthieux (2004).

[13] Putnam's group includes (i) religious or church organizations; (ii) education, arts, music, or cultural activities; and (iii) youth work (e.g., scouts, guides, youth clubs, etc.).

[14] Olson's group includes (i) trade unions; (ii) political parties or groups; and (iii) professional associations.

In the last part of their paper, Knack and Keefer (1997, p. 1277) demonstrate some possible determinants of civic cooperation such as *group membership, income inequality and ethnic polarization, formal institutions for protecting property rights, per capita income, and education rates.*

Helliwell and Putnam, in an article published in 1995 in the *Eastern Economic Journal*, found a strong relationship between social capital and per capita income. According to the two authors, regions with high social capital will have high income levels. Helliwell and Putnam use three indicators to proxy for social capital: (i) an index of civic community; (ii) institutional effectiveness; (iii) citizen satisfaction. The first index is comprised of four indicators: newspaper circulation, the availability of sport and cultural involvements, attendance in referenda and frequency of voting. The second index is compounded by twelve separate elements related to the performance of regional government. Each element is scored, then all components are combined into a single variable to measure the proportional performance of regional government. The last index is calculated by the share of the respondents in some surveys between 1977 and 1988. The question has three alternative solutions: (i) not satisfied; (ii) little satisfied; (iii) rather or very satisfied. In Heliwell and Putnam model,[15] GDP per capita is used to represent 'a composite efficiency measure' and they assume that the convergence between regions is due to inter-regional

---

[15] They used A_{it} to stand for real per capita GDP in region i at the year t, and then the following equation is used for estimating approach: $ln(A_{it}/A_{i,t-1}) = \delta(ln(g_{it} A^*/A_{i,t-1})) + \delta_{it}$. In the equation, δ is the growth rate of productivity and A^* represents income per capita in some leading regions in Italy. After that, the econometric model is $ln(A_{it}/A_{i,t-1}) = a_0 + a_1 \, Civic + a_2 \, lnA_{i,t-n} + \varepsilon_i$, where, $a_0 = \delta ln(A^*)$, $a_1 = \delta ln(g_i) + \delta_i$, $a_2 = -\delta$, and ε_i is the error term, $\varepsilon_i \sim N(0,\sigma^2)$.

transfers of knowledge. Helliwell and Putnam model's assumption is therefore contradicted by Mankiw, Romer, and Wail's (1992) model, as well as that of Barro and Sala-i-Martin's (1991).[16]

In the regression strategy, for the first and second regressions, twenty regions in Italy were covered, while the third regression covered only eighteen regions. Each regression tried to test the effect of civic community, institutional performance, and citizen satisfaction on income at regional levels. All regressions used ordinary least square (OLS) and instrumental variables regressions (IV) as the two main econometric methods. According to these authors, the strong convergence of income per capita was confirmed between Italian regions during the 1960s and 1970s. Interestingly, they found that the regions with more social capital had higher income levels, and the pace of convergence was faster. Divergence among Italian regions appeared after 1983. The authors discovered that *these perceived increases in the effectiveness of regional government also led to greater economic growth*. Indicating the reality of conditional convergence, Helliwell and Putnam conclude that the reversal of convergence in Italian regions after 1983 should be seen as a temporary occurrence. Furthermore, they also point out that their evidence provides a good example of lagging regions in terms of economic convergence due to the differences in the effectiveness of regional government, and following on, the level of social capital.

Based on game theory, Ostrom (1999) provides an example of building social capital and employs results by Wai Fung Lam (1998) as strong

---

[16] In the Barro and Sala-i-Martin (1991) model, the rate of convergence, β, 'depends on the productivity of capital and the willingness to save'. In the model of Mankiw, Romer, and Wail (1992), the rate of convergence depends on rates of saving and population growth.

evidence for supporting her idea. Lam (1998) used data of 150 irrigation systems in Nepal and found some significant evidence for the relationships between 'the physical attributes of irrigation systems, how the systems are governed, and three dependent variables: (a) the maintenance of the physical system; (b) the equity of water delivery; (c) agricultural productivity' (Ostrom, 1999). Ostrom and her colleagues studied irrigation systems in some developing countries and found that farmers should work together constructively in maintaining the irrigation system. According to her, the theory of repeated game can be applied to explain 'how self-interested, calculating individuals can reach cooperative, efficient outcomes in this setting, but the same theory permits inefficient outcomes as well' (Sobel, 2002).

Helliwell (1996) focuses on Asia when trying to establish the source of miracle growth in this area. He carefully compares his work with other scholars such as Mauro (1995) and Inglehart (1994), and also compares his work from 1995 (with Putnam) with his previous paper. *Firstly*, after evaluating the work of Mauro (1995), Helliwell concludes there appears to be a dual relationship between democracy and economic growth: (1) the strong and positive effect of income levels to the presence of democratic institutions; (2) the weak negative or nearly zero correlation among democratic institutions and subsequent economic growth. Moreover, he finds that the opinion of respondents in the survey related to the performance of institutions may not reflect the truth, because business leaders who might be optimistic or pessimistic about the future of economic growth or the quality of institutions are likely to be influenced by other ideas. Therefore, performance of institutions in the form of measuring services supporting business activities is not an ideal model for explaining the income differences between countries. *Secondly*, according to Helliwell, in Putnam (1995, 1996), there exists a negative relationship between social capital stock and modernisation in the USA; however, Mahbubani (1995, p. 107) establishes the reverse correlation in Asian countries: 'The glue that holds Asian societies and families together has not been weakened by modernisation.' In an unpublished paper by Inglehart (1994, cited in Helliwell, 1996), Inglehart highlights significant support for Putnam's arguments. Using the world value survey, Inglehart (1994) establishes that China has the highest index of trust,[17] but that China, Japan, and Korea have lower indices of associational memberships.[18] However, Helliwell (1996) cannot carry out a conclusion for Asian countries due to a lack of data: there are only three Asian countries (China, Japan, and Korea) in the Inglehart dataset.

Granato et al. (1996) incorporate cultural variables in an endogenous growth model to test the effect of culture on economic growth in order to explain cross-country differences. Their endogenous growth model is in the following form: $Y_i = \beta I_{0,i} + \delta_{X_i} + \varepsilon\_i$, where Y_i is the output growth (per capita) of country i, $I_{0,i}$ is the set of economic variables in the initial time period of country i, X is the set of other variables of country i that includes cultural variables. They construct an achievement motivation index[19] based on the data of the world value survey.

---

[17] The index of trust of China is 0.6 compared to 0.5 for the USA, 0.52 for Canada, 0.42 for Japan, and 0.34 for Korea.

[18] This index is 1.94 for the USA, 1.68 for Canada, 1.47 in Korea, 0.83 in China, and 0.49 in Japan.

[19] Achievement motivation index is calculated by following equation: $(A + B) - (C + D)$, where

These authors' empirical results are statistically significant; they find strong evidence that culture has an effect on economic growth in the context of the endogenous growth model. Their evidence is significant in relation to growth theory and offers us a better understanding of economic growth.

4.3.4. Discussion

In this section, a discussion on social capital with respect to its definition and measuring methods is given. This discussion will be presented in three points: (i) argument in concept; (ii) measurement debates; (iii) measuring the relationship between macro levels of social capital and economic growth.

Firstly, let us start with discouragements on behalf of Kenneth Arrow (1999) and Robert Solow (1999). Solow asserts that social capital is 'an attempt to gain conviction from a bad analogy', while Arrow states, 'I would urge abandonment of the metaphor of capital and the term, "social capital"' (Arrow, 1999). Arrow argues that social capital does not have all three characteristics of capital, i.e., (i) extension in time; (ii) deliberate sacrifice for future benefit; (iii) alienability. According to Arrow, social capital has not required sacrifice, though he agrees that social capital shares some temporal aspects with physical capital. He also gives the

---

- A, B: the percentage of people in a particular country that emphasise autonomy and economic achievement.
- C, D: the percentage of people in a particular country that focus on conformity or traditional social norms, such as 'obedience' and 'religious faith'.

point that social capital should not stick to calculation or sacrifice. For example, children decide to join a group without any calculation. Arrow's argument may be true in the case of children; however, in the case of for example businessmen, his argument cannot be used. Businessmen make decisions about joining groups by calculating how their decisions may bring opportunities, such as new contracts, or at least, new and good relationships with others. The calculated mind, therefore, can be assumed to be enhanced during this time period. It reaches the highest point during a particular time in the individual's life cycle, and then reduces due to age, health status, and socio-economic conditions. This idea can be expressed as the following figure:

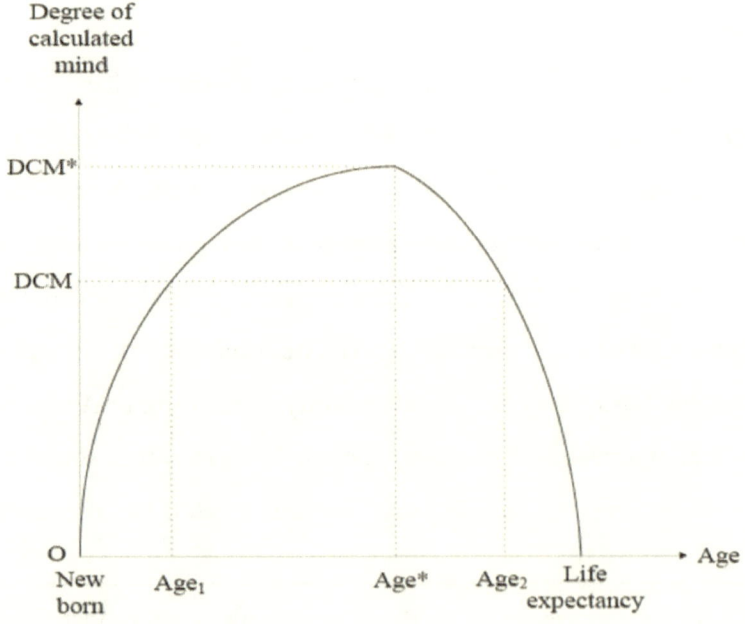

The criticism of the concept of social capital is continued by the research of Woolcock (2001), Davies (2001), and Sixsmith et al. (2001). According to Woolcock (2001), Putnam's definition of social capital neglects the relationship between social capital and power. Moreover, the concept is applied by other researchers for particular countries without consideration of cultural differences. Davies (2001) proposes that Putnam's perception does not include the notion of gender and ethnicity.[20] Additionally, Sixsmith et al. (2001) recommend that numerous studies use secondary data, while very little research employs primary data.

Secondly, social capital is generally understood as a common possession of a group or organisation rather than an individual's asset. Individuals can access these possessions of particular groups or organisations if they are members of them. Hence, the most common measurement of social capital is the participation of individuals such as members of professional organisations or political parties. However, these types of participation actions cover only one or some aspects of social capital, while social capital is considered a multifaceted variable. Côté and Healy (2001) suggest that we should consider a more wide-ranging concept of social capital (as stated above).

The discussion of ways of social capital measurements has led to the particular methods for collecting data in surveys. The lack of empirical practices concerning the calculation of social capital reflects the absence of a complete an encompassing version of questions in the survey questionnaire (Côté and Healy, 2001). The problem in collecting data at the country level causes difficulty in comparing social capital between countries. An attempt at overcoming this problem is the World Value Survey, carried out by sociologists at the University of Michigan. This data, while not perfect,

---

[20] However, it is said that if we put all things into a definition, we may have a long definition that is not easily remembered.

is very useful for the purpose of calculating social capital at the country level in order to make comparisons between countries.

Finally, the effect of institutions on economic growth is widely accepted, but the measurement thereof remains controversial. According to Acemoglu (2009, p. 124), 'we do not currently know enough about the evolution of economic institutions and their impact on economic outcomes to be able to specify and fully estimate structural econometric models'. There is much evidence of the relationship between institutions and economic growth in the history of economy. For example, on their Independence Day, the North and South of Korea have the same starting point. Following several decades, the North has developed slowly, while the South has become a member of the OECD and is now being hailed as a new industrial country. The significant difference between the two regions is institutional structure.[21] In the case of Korea,[22] we do not have to focus on the cultural differences, because North and South share the same history and culture. However, the case of Korea has not led to the conclusion that institutions have a strong effect on economic growth, because this case is not common.

The poor measurements of institutions are caused by the definition of 'institution' itself. The notion of establishment covers a number of different

---

[21] The North applied the model of Soviet Communism and the Chinese way, limiting or destroying the private sector, while the South employed capitalism with incentives, or based on the private sector.

[22] According to Acemoglu (2009, p.126), 'Korea exhibited an unparalleled degree of ethnic, linguistic, cultural, geographic, and economic homogeneity. There are few geographic distinctions between the North and the South, and both share the same disease environment.'

types of social arrangements, laws, regulations, property rights, and so on. All of them have only a few common characteristics, so we cannot use one or two variables as their proxies. In fact, surveys do not include everything related to all elements of social capital. These drawbacks have constrained researchers into constructing a model that cannot clearly and totally explain the relationship between institutions and economic growth. Therefore, the current definition of social capital at the macro level does not reveal the true life of social capital.

4.4. Agricultural development

4.4.1. Economics of agricultural development

In the literature on agricultural development, there exist several ways of explaining the economics of agricultural development. The first is focus on sectoral development, where there is more emphasis on development policies related closely to two sectors: agricultural and non-agricultural areas. The second approach highlights the creation of non-agricultural jobs in rural areas. The researchers try to explain the agricultural development process as the period when agricultural labours decline significantly, and which occurs concurrently with an increase in the number of local non-agricultural employment. Moreover, they include the migration from rural to urban areas as significant evidence of agricultural growth. The third explanation stresses the well-being of agricultural households as a benchmark for the agricultural development of developing countries. These three explanations do not cover all methods of researching agricultural development, but remain the most popular and attract the attention of economists worldwide.

The literature indicates that *the first way* is summarised in Barro and Sala-i-Martin (1995). In the book, there are many modern models that can be used to explain the relationship between agricultural and non-agricultural sectors. Openness levels, government, and macroeconomic indicators have been analysed, but the contribution of these variables to agricultural development has not been directly tested using these models. Timmer (2002), using a panel data of 65 developing countries in the time period 1960 to 1985, found that nearly 20 percent of growth rate in the non-agricultural sector had been contributed by agricultural growth. Naturally, there is no causality relation between agricultural and non-agricultural productivities, but among the two sectors, there exists an effect. According to Timmer, productivity of agricultural and non-agricultural sectors at the same time does not show causation, but the lag of five years of agricultural production, however, demonstrates a more fundamental correlation between them, 'because of the predetermined nature of the lagged variable' (Timmer 2002, p. 1496).

There is a difference between Asian and African countries in terms of agricultural development policy. East and Southeast Asian countries consider agricultural development a high priority sector in their plan, because of its importance in maintaining food security, while African countries employ an urban-biased development policy. Therefore, in African countries, there exists a distortion in the investment environment (Lipton, 1977, 1993). In fact, poverty rates in Sub-Saharan African countries have not declined as impressively as they have in East Asian countries. The rate decreased from 79.8 percent in 1993 to 77.5 percent in 2002 for Sub-Saharan African countries, while the rate of East Asia and Pacific countries reduced from 70.6 percent in 1993 to 45.6 percent in 2002. Clearly, they had the same start; however, there remains a large development gap between them (see table below for more detail).

Table 4.1. Rural poverty rate worldwide 1993–2002

Region	Rural poverty rate ($2.5-a-day poverty line)	
Sub-Saharan Africa	79.8	77.5
South Asia	85.1	83.4
India	89.1	85.6
East Asia Pacific	70.6	45.6
China	72.8	44.6
Middle East and North Africa	23.5	23.5
Europe and Central Asia	16.6	13.6
Latin America and Caribbean	29.6	31.7
Total	**63.3**	**54.4**
Less China	59.6	57.9

Source: World Bank (2008, p. 48)

There are two crucial intellectual developments in agriculture that have been presented since the end of World War II. The first development is the scientific revolution in agriculture, also called the Green Revolution. This revolution has transformed agriculture from subsistence levels to higher levels (Johnson, 1997). The second development is the increasing understanding of the role of the market in economic development and the responsibility of government in fostering this process. Recently, consensus has been reached among agricultural economists that agriculture has been best served by a market-oriented economy (Timmer, 2002).

Todaro (2009) provides a comprehensive summary of this way of thinking. He shows us three broad stages of agricultural development:

(i) peasant subsistence; (ii) mixed family agriculture; (iii) specialised commercial farming. Based on literature related to this issue, Todaro demonstrates that African countries are still in the first stage of agricultural development, while some Asian developing countries are in the second stage and developed countries are in the third stage. According to the author, in the first stage, agricultural households are often resistant to applying innovation to their work. The standard theory illustrates that households will produce at their marginal product or profit-maximising, while in fact, households choose not to use new methods of production because of asymmetric information. Asymmetric information has brought about increasing transaction costs and has placed farmers in front of a price band rather than a single price. In an instance of limited access to credit and insurance, farmers usually choose not to apply innovations.

The second way is comprehensively summarised in a study by Foster and Rosenzweig (2008). The two researchers aimed to answer one of the biggest issues not only in economic development, but also in agricultural development. Historically, in the economic development process, the agricultural sector has always been limited and has generally aligned itself with the expansion of industry or urbanisation. Change in agricultural activities is unclear when it is relegated to the migration of rural people to urban areas. In fact, it is hard to point out whether the reduction of agricultural activities is linked to migration or not (Foster & Rosenzweig, 2008). Consequently, the connection between agricultural development and non-farming activities is not well understood. Generally, agricultural development can maintain long run economic development and agricultural development should be seen as a critical subject for economic development;

however, knowledge about the connection between agricultural development and economic development remains elusive (Acemoglu, 2009).

The third way is summarised in Estudillo and Otsuka (2010). The two researchers, who focus on South-East Asian countries, found that in the case of land frontier and population, the annual growth rate of these countries was 2%; the poverty rate of these countries has decreased gradually along with the shift from agricultural to industrial economies. In these countries, the raising of agricultural productivity had begun prior to transformation of the economy as a whole (Estudillo & Otsuka, 2010). Using datasets at the household level in the Philippines in 1979 and 2004, and in Thailand in 1987 and 2004, Estudillo and Otsuka discovered that the expansion of non-farming activities, and a better chance of agricultural households joining the non-farming labour market to be the main reasons for agricultural development in the two countries. Ravallion and Van de Walle (2008), using datasets of agricultural households in Vietnam from 1993 to 2004, studied the effects of changes in land institutions on the living standard of Vietnamese agricultural households.[23] They demonstrate that the decollectivisation in 1988 started by implementing the first land law reform, principally refraining from 'elite capture' and developed approximately equally on family farms. The important conclusion yielded by their study is that the reform of Vietnam did not focus on maximum economic performance. Instead, the reform has aimed at improving the living standards of Vietnamese people, especially in the case of agricultural households. Based on both economic theory and historical contexts, the

---

[23] In their book, *living standards* refers to household command over commodities, as measured by consumption (the terms *welfare* and *living standards* are used interchangeably) (M Ravallion & D Van de Walle, 2008, p. 2).

two authors suggest that lessons from Vietnam should be considered thoughtfully by China and elsewhere.

4.4.2. Fundamental factors of agricultural development

In literature related to agricultural development, a large number of research papers address investment policy in agricultural development in developing countries. There are both successful and failed instances regarding these policies. In African countries, urban-biased policies have distorted the investment environment of agricultural development, while priority agricultural development policies in some Asian countries have brought these countries significant achievements. Some Asian countries are currently newly industrialised currently. Obviously, agricultural development is both pro-growth and pro-poor.

Mundlak et al. (2002) demonstrate a comprehensive comparison among three Asian countries (Philippines, Indonesia, and Thailand) in the period from 1961 to 1998. They found that in these three countries, which are similar in terms of geography, climate and other shared features, there was a significant difference in agricultural productivity and income. Additionally, the authors reveal that 'factor accumulation played an important role in output growth and that accumulations from policy-driven investments in human capital and public infrastructure were important sources of productivity gains' (Y. Mundlak, D. Larson, & R. Butzer, 2002).

According to Y Mundlak et al. (2002), elasticises of output on inputs in the three countries are quite different. *Firstly*, elasticity of irrigated land in Indonesia is reported at the highest value, 0.46, while the lowest is Thailand at only 0.132. In the case of rain-fed land, the situation is reversed. The highest elasticity is the Philippines, 0.43, while the lowest is Indonesia, 0.23. Total elasticity of land on output, therefore, is highest in Indonesia at 0.69 and the lowest point is Thailand at 0.38. The difference is explained as a consequence of land scarcity in Indonesia being more severe than in the other two countries.

Secondly, the elasticity of fertilisers in these countries was in reasonable agreement, and ranged from 0.06 to 0.084. *Thirdly*, the elasticity of capital was considerably different among the three countries. It was the highest in Thailand, where the elasticity of land was lowest, whereas it was the lowest in Indonesia, where the elasticity of land was highest. *Finally*, the elasticity of labour was relatively low in all sample countries. The contributions of labour to the output in the Philippines, Indonesia, and Thailand were 0.181, 0.198, and 0.144, respectively, lower than 20 percent. In short, the theory framework indicated that the three countries, with similar natural conditions, should have had the same level of output or elasticises of inputs, when in fact, they were different.

In trying to explain this difference, Acemoglu (2009) points out four fundamental factors that should be seen as causes. They are (i) luck; (ii) geography; (iii) culture; (iv) institution. For luck, Acemoglu refers to the 'small uncertainty or heterogeneity between them [countries] that has led to different choices with far-ranging consequences or . . . different selection among multiple equilibria' (Acemoglu, 2009, p. 110). For geography, he refers to the part of 'physical, geography, and ecological environment' where individuals live. Additionally, culture has been linked to 'beliefs, values, and preferences that influence individual economic behaviour'.

Finally, institutions[24] are proximate for 'rules, regulations, laws, and policies that affect economic incentives and thus the incentive to invest in technology, physical capital, and human capital' (D Acemoglu, 2009, p. 111).

4.4.3. Roles of human and social capital in agricultural development

Timmer (2002) argues that the differential level of education between rural and urban areas should be seen as a suitable proxy for the urban-biased policy. According to him, restricting rural investment implies that a reduction in the number of schools and educational facilities have been declined. Timmer also points out that urban-biased development policy has negatively affected rural income, so that the demand for investing in human capital has been reduced. Consequently, rural employees end up in a vicious circle they cannot escape due to a lack of resources.

Additionally, agriculture supports human health by supplying healthy food and nutrition. However, agricultural production and food consumption have increased the risk of contaminated water-related diseases such as malaria and animal diseases (avian flu, brucellosis) (WorldBank, 2008). The relationship between agricultural development and diseases has been reported on worldwide. Poor people tend to be involved in agriculture and their incomes constrain them from having access to healthcare facilities. The case of public health in Italy in the first half of the twentieth century is recounted as significant evidence, and the case of Côte d'Ivoire shows

---

[24] A comprehensive summary on the effect of institutions on agricultural development is presented in chapter 2, where the focus is on historical perspectives.

the strong relationship between malaria and labour productivity. Malaria can be controlled by improving agricultural water systems. In the 1990s, better irrigation and drainage systems in developing countries like the Arab Republic of Egypt, India, and Indonesia halved their malaria cases (WorldBank, 2008).

Fogel (1991) indicates a strong relationship between the nutritional intake of workers and their productivity. Using a comprehensive model linking height, body mass, and mortality, the author demonstrates that an increase in food intake in Britain contributes significantly to the raising of labourers' efficiency and incomes. Fogel states,[25] 'Thus, in combination, bringing the ultra poor into the labor force and raising the energy available for work by those in the labor force explains about 30 percent of the British growth in per capita incomes over the past two centuries' (p. 63). The correlation between nutritional intake and agricultural productivity can be used to explain the miracle in East Asian countries during the twentieth century (WorldBank, 1993). In an effort to measure the effect of nutritional intake on labourers' productivity, Nadav (1996) extended the model of Mankiw et al. (1992) to include nutritional capital for a sample of 97 countries in the period 1961–1981. He found a strong correlation between nutritional intake and economic growth.[26] Moreover, Nadav (1996) concludes that nutritional intake remains significant for explaining economic growth, while the variables computing the rate of schooling and labour growth are inconsequential.

---

[25] Cited in Timmer (2002).

[26] The coefficients of ln (nutrition) are 0.960 and 0.933.

4.4.4. Policy implications

The major objective of agricultural development is the progress of implementing public policy for boosting the agricultural sector to achieve long-term economic growth. There have been many concerns about agricultural development policy in the past; currently, however, this issue appears to have lost traction among researchers and the governments of developing countries. Urban-biased development policy has brought about the concern of food security worldwide and low development levels and living standards in agricultural and rural areas. Todaro (2009) provides a complete summary of this issue and supplies a list of potential policies for developing countries to gain a high level of success in their agricultural development strategies.

The first policy he suggests deals with **technology and innovation**. Todaro (2009) acknowledges that in developing countries, it is sometimes difficult to apply technological innovation for two major reasons.

Initially, introducing new mechanics into agriculture will absorb agricultural labours and can raise the ratio of output per worker. Unfortunately, in low-developed countries (LDC), particularly in rural areas, capital is inadequate and labourers are abundant. If the governments of LDCs employ new technical innovation for raising agricultural productivity, they will have to face the more severe issue of rural unemployment. Moreover, modern technological innovation usually excludes female workers in the agricultural sector, so the gap between male-female in terms of productivity will be widened, with serious consequences (Binswanger, 1986; Binswanger & Pingali, 1988).

Additionally, biological, irrigation, and chemical innovations are difficult to implement in developing countries due to their agricultural development problems. LDCs' agriculture is usually land-augmented, which means that farmers focus on their particular space of land or improve the quality of land in order to raise production per hectare. Clearly, applying modern technology in agriculture in these instances can boost agricultural productivity, as seen in the Green Revolution in Asia during the previous century. In fact, the diffusion of the Green Revolution is present only in Asian countries. Agricultural systems in African countries require innovation, such as the Green Revolution; however, extension of agriculture is still not happening in Africa (Anderson & Feder, 2007; Huffman & Orazem, 2007; Todaro & Smith, 2009).

The second policy recommendation relates to **institutions and prices**. Todaro (2009) provides an excellent summary in terms of public policy related to this important issue. As suggested by Acemoglu (2009), an institution with a fundamental factor created unequal with respect to the output. In this case, the propositions of the two prominent professors are suitable and are suitable for explaining the current situation in developing countries.

The Green Revolution has provided a wide range of options for farmers who require irrigation systems and chemicals; however, not all farmers are able to achieve the outcomes resulting from innovations. The reasons for this restriction are institutional failure and pricing system. In the context of developing countries, only wealthy farmers have enough resources to apply new innovations, which give them a competitive advantage when compared with smaller landowners. In addition, local authorities usually provide these tools to rich farmers because they know that poor farmers cannot afford them (Conning & Udry, 2007; Todaro & Smith, 2009). Consequently, the effect of the Green Revolution in South Asia and Mexico is constrained and diffusion of the Green Revolution does not reach African countries.

At the micro level, the result of the above discussed facts is the widening of the gap between rich and poor farmers. This gap has significantly affected the opportunities of access to rural financial markets by the poor, while rich farmers are not as affected as poorer farmers. Large-scale farmers have been able to access low-interest loans from the government, while poor cultivators find themselves paying interest to moneylenders. Poor farmers cannot borrow money from the bank and therefore have to borrow money from a third party with higher interest rates than banks. This ties poor farmers to poverty in the long run, which they cannot escape without support from the government.

The third policy Todaro advises is a strategy for **adapting to new opportunities and constraints**. Todaro (2009) believes that agricultural workers today have to face both opportunities and constraints to attain new achievements. According to him, the best opportunities for farmers are not improved crops' productivity. Instead, farmers should focus on higher value-added horticulture and aquaculture. These approaches will have higher value in exporting and boosting urban areas. In order to implement this strategy, small farmers need to be supported by the government. The World Bank (2008, p. 338) concludes, 'Smallholders can bargain better as a group than as individuals. So a high priority is to facilitate collective action through producer organizations to reach scale in marketing and bargain for better prices.' However, modern farmers face a growing threat in the form of global warming and climate change. Sub-Saharan Africa and South Asia are expected to endure the most harm from climate change and global warming. Both developed and developing countries should realise their responsibility within this context. Developed countries contribute too much CO_2 emissions, while developing countries eliminate too much forest area for expanding cultivated land (Mendelsohn, 2007; WorldBank, 2008). Moreover, the mismanagement of agricultural development has brought about the high cost of fixing this issue, as agricultural development without effective management damages ecosystems and biodiversity.

4.5. Conclusion

This chapter conducted a literature review on human capital, social capital, and agricultural development from an economic perspective. Ideas on human and social capital of sociologists and economists were analysed and classified into three categories: definitions, ways of measuring, and empirical results. Based on these examinations, some discussions were presented to compare the disadvantages and advantages of particular methods. This exercise is useful for choosing an appropriate method to construct human and social capital indices for the following chapters. Additionally, the links between literature on human and social capital and agricultural development were explored to explain the process of agricultural transformation in a new way.

The analysis of current literature suggests that human capital and social capital have contributed considerably to agricultural development. Indeed, human capital formations have a determinant role in transforming agriculture from low to high levels. Social capital formations have a supportive role in transforming agriculture, but the effects of this type of capital have reported problems. Generally, human capital has positively affected productivity, while in some countries social networks have positively influenced the development process; in other countries, social connections have not supported the path of development. Therefore, the initial conclusion here is that human capital should be developed expansively, whereas social capital should be developed selectively.

Chapter 5

THE DETERMINANTS OF HEALTH INVESTMENT TO AGRICULTURAL DEVELOPMENT

5.1. Introduction

This chapter provides an analysis of the effect of health investment on the income of agricultural households in Vietnam over the period 2002–2004. The relationship between health and income is an active subject, not only in microeconomics or health economics, but also in population and health sciences. Recent and on-going research emphasises the complex relationship underlying the joint evolution of health and socio-economic achievement over the life course of people. Other works focuses on the isolation of some causal effect of different health variables on indicators of socio-economic status. Additionally, researchers also target the effect of individual, family, and community resources on health.

Studies have highlighted the ways in which investment in early life health—including foetal health and child health—potentially affect health, income, and well-being over the individual's entire life course. These studies draw attention to the importance of understanding investment strategy in health from life's conception to its end. Studies further indicate a key role for genes and interaction with the environment in influencing the strategy of investment in health throughout life. This emerging body of research promises to contribute to an understanding of the relationship between investment in health and income growth.

At the macroeconomic level, the contributions of health on economic growth and labour productivity have been the main subject of numerous recent research papers and projects. One of the key arguments is that better health can enhance worker productivity and thereby motivate economic growth. Bloom, Canning, and Sevilla (2004) estimated the effect of health on economic growth by using a production function containing two main elements of human capital: work experience and health. They found that 'good health has a positive, sizable, and statistically significant effect on aggregate output even when we control for experience of the workforce' (Bloom, Canning, & Sevilla, 2004, p. 1). These authors used a sample of 104 developed and developing countries and concluded that 'a one-year improvement in a population's life expectancy contributes to an increase of 4% in output' (Bloom et al., 2004, p. 11). In addition, Sala-i-Martin, Doppelhofer, and Miller (2004) remark that life expectancy has a positive effect on economic growth, while the prevalence of malaria has a negative effect on economic growth. In support of the second finding of these authors, the United Nations (2005) suggests that malaria slows economic growth in Sub-Saharan African countries annually by 1.3%. Jamison, Lau, and Wang (2004) restrict their health measure to the survival rate of males aged 15 to 60 and estimate that improvement in health may have contributed to 11% growth from 1965 to 1990 in 53 countries. According to the authors, the effect is higher in particular countries, such as Bolivia, Honduras, and Thailand. In support of this result, Gyimah-Brempong and Wilson (2004) found that health factors add between 22% and 30% to the rate of economic growth in Sub-Saharan African countries.

The primary goal of this chapter is to measure the rate of return at the micro level on investment in health capital during the period 2002–2004. The author does not focus on the effect of health outcomes on household incomes or allocation of resources to health or other sectors. Due to a lack of data, this chapter also does not analyse the effect of life expectancy on households' income at individual level. Analysis of healthcare financing, production, and distribution of healthcare services are therefore not the goals of this chapter.

5.2. Literature review

Over the past two decades, there have been numerous research papers on investment in human capital. Human capital has two major elements: human education and health capital. Wage function estimations have offered a popular way of estimating the rate of return to education in almost every country. However, whereas researchers have focused on education, returns to investment in human health capital remain under-researched. The relationship between human health capital and income at the micro level is complex and not easy to explain. This section provides an analysis of current and on-going related economic literature. Based on the analysis, the theoretical and empirical framework will be built in the next section.

5.2.1. Human health capital investment

5.2.1.1. The definition of human health capital

Human health capital is a complicated term and it is difficult to provide a definition that covers all dimensions of human health capital. Schultz (2005) asserts that 'there is no consensus among health specialists on how to conceptualise and measure health status at the individual level' (Schultz, 2005, p. 1). This is because 'health' has many dimensions and each element has its own characteristics; therefore, they need to be measured in specific ways.

The objective of defining human health capital is to estimate the effect of health indicators on labour productivity, which is approximated by labour income. Family background, culture, religion, and community membership can all affect human health capital and are referred to as reproducible human health capital.

Dolan (2000) defines health outcome as 'the benefit that a patient derives from a particular health care intervention [and which] is defined according to enhanced quality and/or length of life' (Dolan, 2000, p. 1725). Using this definition, Dolan employs cost-utility analysis instead of cost-benefit analysis to examine health outcomes and the decision to invest in the health of individuals.

Health is central to human wealth and a prerequisite for high productivity labour. In the literature, good health leads to good education and a better life (see Strauss & Duncan, 1998). From this definition, health can be seen as one of the vital components for growth and development, i.e., an input for aggregate production function.

5.2.1.2. The way of measuring health capital

There are several ways of measuring health capital at individual and aggregate levels. Strauss and Thomas (1998) analysed two methods: (1) height of adults and (2) body mass index (BMI).[27] For the first method, the authors believe that height can effectively link human health capital investment during childhood with education level. A person with high levels of health investment will grow taller and gain a better education. Therefore, height is a measurement that can express not only the quantity of education (one main dimension of human capital) but also the quality of schooling. The researchers also believe that human capital investment in childhood will be reflected by the good health status of an adult. The second method can measure variations in health and nutritional status in the human life cycle. According to the authors, 'Current BMI partly reflects previous health and human capital investments, and so a correlation between BMI and productivity may be capturing the influence of those prior investments' (J. Strauss & D. Thomas, 1998, p. 774).

Another method of measuring human health capital is a general health status (GHS) survey, which is conducted by governments or private institutions. Commonly, a general health status survey is usually part of a household survey. Respondents are often asked to rate their health status according to categories of four to six, ranging from poorest to excellent health. This method cannot provide accurate results, because the answers are limited and respondents sometimes do not know how to rate their health. In addition, the term 'good health' can be understood differently from person to person. Results may therefore not be reliable for

---

[27] BMI = Weight (in kilograms)/Height^2 (in meters).

public policy making. Consequently, researchers do not use this type of measurement directly and instead use correlations between GHS indices and rates of morbidity and mortality (see Ware, Davies-Avery, & Donald, 1978). Other researchers incorporate information on prior health care, socio-economic characteristics such as values, family background, beliefs, and wage or income within GHS measures, which makes the results more reliable and more accurately reflects levels of human health capital.

Increasing the relevance of GHS measures proved very useful to researchers who focused on two sets of health status indicators: (i) self-reported morbidity, illness and 'normal' activity and self-reported physical functioning; and (ii) nutrition-based indicators. These two ways of calculating human health capital were reported by Strauss and Thomas (1998). According to the authors, the survey data that has self-reported answers in developing countries is usually difficult to interpret. For example, Schultz and Tansel (1997) found that ill-health is correlated positively with education levels in Ghana and Côte d'Ivoire. Hill and Mamdani (1989) reported a negative relation between days of ill-health and education level, while Schultz and Tansel (1997) found a positive relation. Self-reported physical functioning is related closely to physical health problems, but it is claimed that these indicators may not be appropriate in a study of health and labour outcomes.

In the literature summarised by Strauss and Thomas (1998), there is consensus that the energy output of workers is not measureable, so researchers focus instead on nutrition intake to find a link between nutrition intake and labour outcomes. Energy output is usually calculated using mainly two methods. The first is to estimate calorie availability as proxy for food consumption. This method has some advantages, such as ease of data collection and measurement, as it is sufficient to have access to data on households' food expenditure. The first method, however, also contains several drawbacks. For example, food is assumed to be consumed (not wasted). Another common measurement error stems from instances where people consume more food on specific days, such as New Year and family anniversary days, than on normal days. The second method surveys nutrition intake during a 24-hour period. Respondents are asked about their meals during the last 24 hours, including food and meals received or purchased away from home. This method overcomes the disadvantages of the first method and provides a more accurate measurement for the energy output of workers.

These methods of calculating human health capital imply a correlation between human health capital investment and personal income at individual or aggregated level.

5.2.1.3. The relationship between health capital and productivity

Measuring the contribution of health capital to labour productivity has attracted many researchers who have illustrated a variety of ways for calculating this relationship. In a series of papers from 1992 to 1994, Robert Fogel measured the effect of workers' height to individuals' standard of living in the long-term. Fogel studied adults in four countries: the United States of America (USA), Brazil, Côte d'Ivoire, and Vietnam, and found that the differences in the heights of adults in these countries could explain their diverse incomes. In addition, this information can be extended for analysing the wealth and health status differences between the four countries.

At the household level, many researchers point out the correlation between income and health status. Jere Behrman and Anil Deolalikar

(1988) and Strauss and Thomas (1995) agree that the poorest household will have the worst health status. They found that in Brazil 'a 1 percent increase in height is associated with an almost 8 percent increase in wages' (J. Strauss & D Thomas, 1998, p. 772), while in the USA, a 1 percent increase in height leads to a 1 percent raise in salaries. According to these authors, taller men are associated with higher educational attainment in both countries. Moreover, the better health status worker tends to complete more efficient, longer hours, and be more productive than the worker who has lower health status. Incorporating people's heights as a proxy for health status, Thomas and Strauss (1997) found that shorter men were more often unemployed than taller men.

To capture the effects of health on productivity, which is proxied by wage, Strauss and Thomas (1998) used two functions: the first function was a 'generic health production function for an individual' (J. Strauss & D Thomas, 1998, p. 775); the second function was real wage, with an assumption that 'an individual's real wage is equal to his marginal product' (J. Strauss & D Thomas, 1998, p. 776). The health production function is presented as follows: $H = H(N,L;A,B',D,\mu,e_h)$ where H is an array of measured health outcomes, N is a vector of health inputs, L is labour supply, A is socio-demographic characteristics, B' is family background, μ is 'the inherent healthiness of the individual' (J. Strauss & D. Thomas, 1998, p. 776) and e_h is measurement error. The real wage function is shown as: $\omega = \omega(H;A,S,B,I,\alpha,e_\omega)$ where ω is the (log) real wage of an individual, S is schooling human capital, B is family background (including parents' years of schooling), I is the infrastructure investment of the local community, α is unobservable factors, and e_ω is measurement error, which captures stochastic fluctuations in the real wage of a person.

The effects of human health capital on labour productivity have been reported in the history of economic literature. Research outcomes are quite complex and there is no unanimous agreement on these outcomes. Nonetheless, the relationship between height and labour outcomes reported above has sufficient consensus amongst researchers. In contrast, the link between ill health and wage is more ambiguous. In an experimental study in Tanzania, Fenwich and Figenschou (1972) found that workers who have suffered from schitosomiasis have lower productivity than other workers who do not have this disease, while Gateff et al. (1971), cited in Strauss and Thomas (1998), found no such difference for the same disease in Cameroon. The difference between the two experiments suggests that we may not be able to apply this method for analysing contributions of human health capital to labour productivity, due to the endogeneity of health. Another disadvantage of experimental studies is that their results do not have great generalisation power due to small and biased samples. It is therefore argued that non-experimental methods offer a better choice for analysing the link between human health capital and labour productivity.

In non-experimental studies, researchers often use data from household surveys, relying on self-reported data from respondents. This way of proceeding is exposed to the health measurement errors discussed above. Similar to experimental research, non-experimental investigations have produced mixed results. Schultz and Tansel (1997) affirm that ill-health Ivorian male workers earn lower wages than good health workers, while Pitt and Rosenzweig (1986) conclude that in Indonesia's rural area, farmers' productivity has not been affected by ill-health farmers. In the case of Indonesia, the result can be attributed to social networks: if farmers cannot work, they will ask their neighbours or relatives to help them. The situation in Indonesia is similar to the case of the Vietnamese agricultural sector.

In addition, there is no consistent answer to the problem of measuring the relationship between illness and labour force participation. Some research found that health has a substantial effect on measures of labour participation. For example, Luft (1975, 1978) found that the health effect is larger for blacks than whites and greater for women than men. Similarly, Berkowitz et al. (1976) found a comparable black-white outcome. On the other hand, Grossman and Benham (1974) failed to isolate health as a significant determinant of incomes. All four studies relied on the standard economic theory of labour-leisure choices. The different results among them can be interpreted as a consequence of the way in which these studies measure health—they all used disability as a proxy for people's health status. Since disability was based on self-reports, it may be an over- or under-estimate of the effect of health on economic activities.

5.2.2. Agricultural development

Throughout economic history, most societies developed their agriculture from the hunter-gatherer period onwards prior to developing into industrial economies. This switch is explained using two theories. The first relies on the theory of diminishing return, while the other is based on the theory of increasing return with the diffusion of technology and knowledge. In the first theory, North and Thomas (1977) show evidence that the transformation is caused by diminishing return, a characteristic of the hunter-gatherer economy. Similarly, Smith (1975) presents a model where the hunter-gatherer economy's diminishing return is due to overexploitation of natural resources. Locay (1989) develops a model that connects the fertility decision with agricultural engagement. In his model, population growth has led to increasing opportunities for engaging in agricultural activities, and then lowering the costs of child-rearing. In terms of the second theory, Myers and Marceau (2006) believe that technological progress is an important factor for switching from hunter-gatherer to agricultural societies. According to them, technological progress puts an unbalanced force on the cooperative activities of hunting and gathering. Then, it destroys hunting and gathering activities. Consequently, a switch to agriculture occurred.

In addition, other research is based on the idea that technological progress depends on the density of population, environmental conditions, and cultural factors. These studies explain more clearly the relationship between hunting, gathering, agriculture, technological progress, and population growth. Olsson and Hibbs (2005) demonstrate that geographical and environmental conditions have significantly affected the rate of technology diffusion by means of expanding the frontiers of experimental activities. Moreover, Dow et al. (2005) provide strong evidence that climate reversals have played a crucial part in transforming the economy from hunting to agriculture. At the micro scale, Morand (2002) successfully explains the connection between human capital accumulation, interfamilial household transfer, and hunting and agriculture. Weisdorf (2005) argues that agriculture produces more food than a hunter-gatherer economy and provides a more efficient environment for increasing learning activities. According to him, this advantage is the main reason for the switch to agriculture.

Todaro (2009) proposes three broad stages of agricultural development based on the work of Weitz (1971). The first stage is the lowest productivity period, its main characteristic being that peasant farms dominate agriculture, such as in African countries. The second stage is the period of mixed family agriculture. In this phase, farmers sell a significant part of their harvest to the commercial market, as in Asian countries. The third stage is high-productivity specialised farms. This phase belongs to developed countries or highly urbanised developing countries.

Agricultural development in Asian countries is found in the mixed-market economies. Therefore, their main character of development is 'gradual but sustained transition from subsistence to diversified and specialised production'. Transforming traditional to modern agriculture requires more than reorganising the structure of the economy or applying new technology. Indeed, transformation requires fundamental changes in the social, political, and institutional structures of rural communities. Without these changes, agricultural development cannot achieve its goals, and the gap between large land owners and landless farmers remains or expands.

5.3. A basic conceptual framework

5.3.1. Households production function

It is assumed that a farmer's income comes from two main sources: level of input usage and the technical tools used for working, such as irrigation system, agricultural machineries, insecticides, and fertilisers. The production function is assumed to be of the Cobb-Douglass form:

$$Y_{i,t} = A_{i,t} K_{i,t}^{\alpha} \left(HealthL \right)_{i,t}^{\beta} e^{\cdot \, _1 s_{i,t} +, \, _2 exp_{i,t} +, \, _3 exp_{i,t}^2 +, \, _4 h_{j,t} +, \, _m z'_{i,t}}, \quad (5.1)$$

where Y_{it} is the income of agricultural household i at time t; A represents the stock of technical tools used by household i at time t; K is the stock of physical capital of the household; *Health* is the stock of health capital embodied inside farmers and L is labour; thus, (*HealthL*) stands for good health worker. Human capital consists of three main elements: years of schooling (s), work experience (*exp*), and health ($h_{j,t}$). At an aggregated level, life expectancy has been used as a proxy of $h_{j,t}$. In this instance, life expectancy is used as a proxy of $h_{j,t}$ in order to capture the effect of general health conditions of province j at time t.

Bloom, Canning, and Sevilla (2004) define working experience '[a]s simply the amount of time spent in the labor force' (Bloom et al., 2004, p. 7). Thus, work experience is 'average age minus average years of schooling minus the age which schooling starts, which we uniformly assume to be six' (Bloom et al., 2004, p. 7). In this instance, I have applied this method, but did not use the average level. Instead, I used the individual level, which represents individual years of schooling and age.

z'_t represents other factors that can affect health investment behaviour, such as ethnicity, cultural background, geographical differences, married status or gender, and family background, and m is the number of effects. Function 5.1 is chosen due to the fact that when we take the log form it is compatible with the microeconomic income equation form. Income depends on the level of schooling, experience, experience squared and health status. In addition, the Vietnamese agricultural sector is dominated by households and each household does its jobs itself due the fragmentation of their agricultural land. Consequently, it is more convenient to assume that L is equal to 1, with Function 5.1 written as follows:

$$Y_{i,t} = AK_{i,t}^{\alpha} Health_{i,t}^{\beta} e^{\theta_1 s_{i,t} + \theta_2 \, exp_{i,t} + \theta_3 \, exp_{i,t}^2 + \theta_4 h_{i,t} + \theta_m z'_{i,t}}, \quad (5.2)$$

taking log both sides of (5.2) and define y_{it}, k_{it}, and $health_{it}$ are the logs of $Y_{i,t}$, $K_{i,t}$, and $Health_{i,t}$, respectively, yields:

$$y_{i,t} = a_{i,t} + \alpha k_{i,t} + \beta health_{i,t} + \theta_1 s_{i,t} + \theta_2 exp_{i,t} + \theta_3 exp_{i,t}^2 + \theta_4 h_{i,t} + \theta_m z_{i,t}', \quad (5.3)$$

The neoclassical growth model commonly assumes that a_{it} cannot be observed and it usually appears in the error term. On the other hand, in the context of the endogenous growth model, a_{it} can be seen as a combination of fundamental and secondary innovation. Fundamental innovation refers to the research and development sector, while secondary innovation represents learning by doing (LBD). Consequently, technological progress is captured by a_{it}, a function of fundamental and secondary innovation, and this function has the following form:

$$\frac{\dot{A}_{it}}{A_{it}} = F\left(\vartheta^r R^r, LBD\right), \quad (5.4)$$

Equation 5.4 has to satisfy two assumptions:

(i) $F = 0$ when $R^r = 0$ and $LBD = 0$,
(ii) F is strictly increasing and concave

In equation 5.4, ϑ^r is each researcher's Poisson arrival rate of fundamental innovation in agricultural science, an exogenous parameter, while R^r denotes the mass of researchers related to agriculture. Therefore, $\vartheta^r R^r$ is the flow of new products. When a research product is established, we do not know the quality of the product. The quality is only known in the process of applying the work of agricultural workers in agriculture. If we denote $quality_j$ as the quality of product j, then $\frac{dquality_j}{d_j} = LBD$. In the context of Vietnamese agriculture, fundamental innovations are provided by government institutions, and farmers can have the latest products with support from the government. In addition, fundamental innovations are non-rival goods, and farmers can access them without limitations. In other words, fundamental innovations in the Vietnamese agricultural sector satisfy Pareto efficiency. Therefore, we can assume that $\vartheta^r H^r$ is constant for all farmers, and equation 5.4 can be rewritten as follows:

$$\frac{\dot{A}_{it}}{A_{it}} = F(LBD), \quad (5.5)$$

If we denote ϑ^d to be the productivity of learning by doing and x_j the labour requirement for applying new invention to farm households, LBD can be measured as follows:

$$quality_0 = 0, \frac{dquality_j}{d_j} = LBD = \int_0^\infty \vartheta^d x_j^{1-v} dj, \text{ with } 0 < v < 1, j > 0. \quad (5.6)$$

We can use Figure 5.1 below to express equation 5.6.

LBD at the time t relates to LBD at the time t-1 due to the flow of experienced knowledge. The way of doing and thinking at the present time depend on farmers past time. Farmers cannot change their agricultural behaviours annually and have to keep their working routines the same in the long run. Therefore, we can assume that

$$LBD_{i,t} = \omega LBD_{i,t-t_1} + \varepsilon_{i,t}, \quad (5.7)$$

$0 < \omega < 1$, and ω is being different from one farmer to other farmer. $\omega LBD_{i,t-t_1}$ represents a part of old knowledge embodied inside farmers after the working period and $\varepsilon_{i,t}$ represents proxies for new knowledge. Old knowledge has been embodied inside the worker since the last period of working, while 'new knowledge' is accounted as an error term, as new knowledge cannot be observed directly.

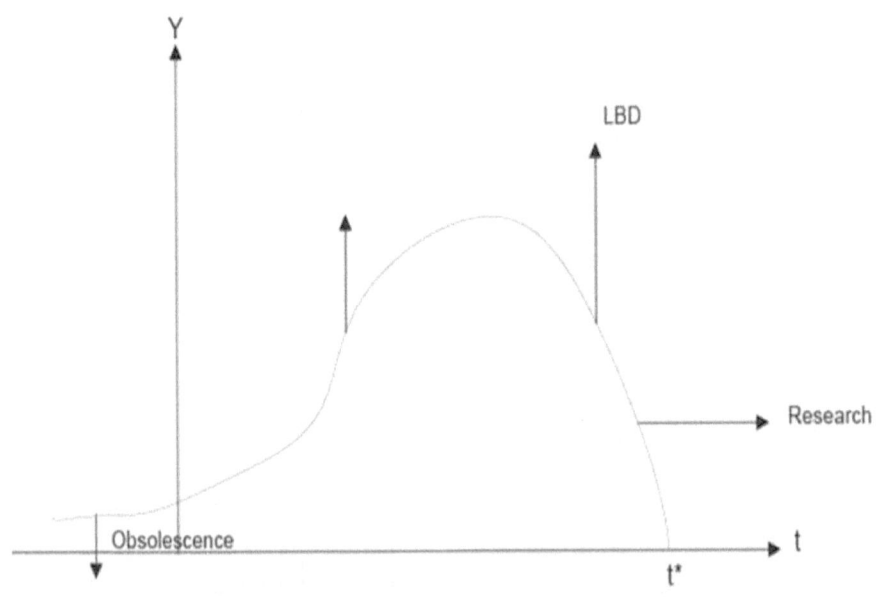

Figure 5.1: The profile of output across line of different vintage, at date *t*

Source: Aghion & Howitt (1998)

The above figure was not originally drawn for agriculture but for industry. Thus, the meaning changes slightly when we apply it to agriculture, especially within the context of Vietnamese agriculture. First, agricultural research and development products such as insecticides, fertilisers, and seeds are the results of biology and chemistry, and are provided by government research institutions. The outcomes are delivered to farmers via local agriculture incentive offices. Farmers can then use the latest inventions for their crops.

Government research institutions can be seen as a force shifting the profile to the right. Secondly, whether research outcomes have good quality or not depends on the process of applying new technology or ways of harvesting and working of farmers. If farmers have good health and experienced knowledge, they will be effective in following the directions of researchers and observing the effect of research outcomes on their crop in the long term. By doing so, farmers learn new knowledge from researchers and can report to researchers some points that can help invent new products. Through learning by doing, farmers improve the quality of new products. Finally, Figure 5.1 implies that farmers and government researchers do connect to improve agricultural and research products. This point is very important for policymakers.

In equation (5.2), $K_{i,t} = sY_{i,t}$, where s is an exogenous fraction of income, it is assumed that physical capital depreciates at an exogenous rate δ. Thus,

$$\dot{K}_t = sY_t - \delta K_t, \quad (5.8)$$

$Health_{i,t}$ is defined for human health capital. Farmers have an endowment of human health capital, $Health_{i,0}$ from the time they first join the labour force. Over their working life cycle, farmers' health will first go up, reaching a highest point of health status and then go down until they cannot work anymore. To maintain human health capital, farmers will spend a part of their income buying health care services in the hope that their health will be good enough to sustain working. Good health workers always have higher productivity than workers who have a lower health status. In fact, the health capital stock of farmers can be presented as a function of time, $Health_{i,t} = f(t)$ and $f(t)$ shaped like the figure below:

Figure 5.2: The timeline of health status of the worker

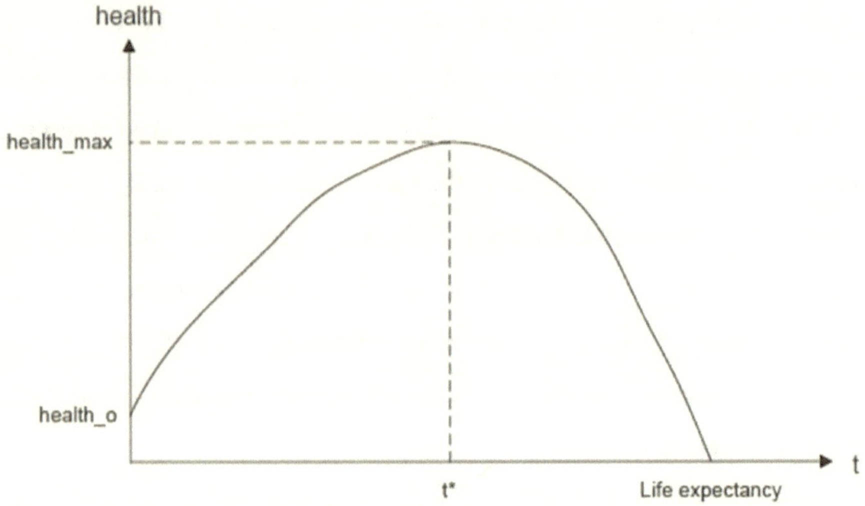

Agriculture workers will have the highest health status at age t*, but there is no way of saying that every worker will have the same year t*. In the agricultural sector there is no retirement age, so it is assumed that farmers can work until they die. The time of death is different from person to person; therefore, life expectancy can be used to express the maximum year of living. Working life is measured by the difference between life expectancy and the age that a farmer joins the labour force (age*), so working life = life expectancy – age*. According to labour law in Vietnam, a person can join the labour force when his age is 15, while in the agricultural sector, the age of joining the workforce may be earlier. To simplify analysis, we assume that age* is 15. From 15 years old to t* years old, health status depreciates with positive sign, and from t* years old to life expectancy age, the depreciation rate has negative sign. Consequently, in the first part of working life, health capital investments improve farmers' health status, and in the second part health capital investments maintain farmers' health status. Therefore,

$$Health_{i,t} = Health_{i,0} e^{\delta_t^H t}, \quad \text{then } Health_{i,t} = \delta_t^H Health_{i,t} \quad (5.9)$$

In addition, if we denote $K^\alpha = v_k \times K^{\frac{\alpha}{\alpha+2}}$ and $Health^2 = v_h \times Health^{\frac{2}{\alpha+2}}$ where v_k and v_h are constants, then a substitute to 5.1 yields:

$$Y_{i,t} = A_{i,t} v_k v_h K^k Health^{1-k} e^{\theta_1 s_{i,t} + \theta_2 \exp_{i,t} + \theta_3 \exp_{i,t}^2 + \theta_4 h_{i,t} + \theta_m z'_{i,t}}, \quad (5.10)$$

with $k = \dfrac{\alpha}{\alpha+\beta}$ and $k = \dfrac{\beta}{\alpha+\beta}$.

Rearranging 5.10, we thus have:

$$Y_{i,t} = A_{i,t} v_k v_h K_{i,t} \left(\dfrac{Health_{i,t}}{K_{i,t}} \right)^{1-k} e^{\theta_1 s_{i,t} + \theta_2 \exp_{i,t} + \theta_3 \exp_{i,t}^2 + \theta_4 h_{i,t} + \theta_m z'_{i,t}}, \quad (5.11)$$

Equation 5.11 implies that if we keep $\dfrac{Health}{K}$ as a constant, the output-capital ratio of households is also constant. Thus, the difference of household incomes depends on $A_{i,t}$. Moreover, $A_{i,t}$ is defined by equation 5.5, so skills obtained from learning by doing have a crucial role for improving agricultural household incomes, and enhancing their living standards. In turn, better living standards will lead households to concentrate on their health and training. Consequently, they can adapt to the urbanised process in Vietnam.

5.3.2. Households preferences

The household utility function is assumed as follows:

$$U_{i,t} = \frac{((C_{i,t}^H)^q (C_{i,t}^{NH})^{1-q})^{1-b}}{1-b}, q \in (0,1), b \neq 1, \quad (5.12)$$

where:
- $C_{i,t}^H$ is health consumption of household i at the time t.
- $C_{i,t}^{NH}$ is non-health consumption of household i at the time t.

Identifying v_c as the fraction of health expenditure on total consumption of the household, we have

$$C_{i,t} = C_{i,t}^H + C_{i,t}^{NH} = v_c C_{i,t} + (1-v_c)C_{i,t}, \quad (5.13)$$

Substituting 5.13 to 5.12, we get

$$U_{i,t} = \mathcal{B}\frac{(C_{i,t})^{1-b}}{1-b}, \text{ with } \mathcal{B} = v_c^{q(1-b)}(1-v_c)^{(1-q)(1-b)} \quad (5.14)$$

If it is assumed that households have an initial endowment, referred to as E_0, then the household i permanent income in the period of T can have the following form:

$$E_0 + \sum_{t=1}^{T} Y_{i,t}$$

Therefore, budget constraints for health expenditure are

$$\sum_{t=1}^{T} C_{i,t}^{H} \leq E_0 + \sum_{t=1}^{T} Y_{i,t} - \sum_{t=1}^{T} C_{i,t}^{NH}$$

Denoting $Y_{i,t}^{H} = Y_{i,t} - C_{i,t}^{NH}$ as the potential part household incomes will spend on health, health expenditure constraints can be rewritten as

$$\sum_{t=1}^{T} C_{i,t}^{H} \leq E_0 + \sum_{t=1}^{T} Y_{i,t}^{H} \quad (5.15)$$

Equation 5.15 implies that health consumption should never exceed a household's income and its initial wealth. In the case of $C_{i,t}^{H} > Y_{i,t}^{H}$, families can reduce $C_{i,t}^{NH}$ and use E_0 to borrow money from the bank, their friends or relatives. If they borrow from the bank, the loan will never exceed E_0, while they can borrow over E_0 from their friends or relatives. In short, a household's optimal problem is maximising its utility as per equation 5.14, subject to the budget constraints in equation 5.15.

5.4. Methodology

For estimation purposes, the growth equation is calculated from equation 5.3 by differentiating both sides and yields:

$$\Delta y_{i,t} = \Delta a_{i,t} + \alpha \Delta k_{i,t} + \beta \Delta health_{i,t} + \theta_1 \Delta s_{i,t} + \theta_2 \Delta exp_{i,t} + \theta_3 \Delta exp_{i,t}^2 + \theta_4 \Delta h_{i,t} + \theta_m z_{it}' \quad (5.16)$$

Substituting equations 5.5 and 5.7 to equation 5.16, then rearranging, we have

$$v\Delta y_{i,t} = \frac{\alpha}{\omega}\Delta k_{i,t} + \frac{\beta}{\omega}\Delta health_{i,t} + \frac{\theta_1}{\omega}\Delta s_{i,t} + \frac{\theta_2}{\omega}\Delta exp_{i,t} + \frac{\theta_3}{\omega}\Delta exp^2_{i,t} + \frac{\theta_4}{\omega}\Delta h_{i,t} + \frac{\theta_m}{\omega}z'_{i,t}$$
$$+ \frac{1-\omega}{\omega}(\alpha k_{i,t_1} + \beta health_{i,t_1} + \theta_1 s_{i,t_1} + \theta_2 exp_{i,t_1} + \theta_3 exp^2_{i,t_1} + \theta_4 h_{i,t_1}$$
$$+ \theta_m z'_{i,t_1} - y_{i,t}) + \frac{\varepsilon_{i,t}}{\omega}$$

Equation 5.11 shows that growth in household income can be decomposed into four elements: (i) vintage knowledge; (ii) growth of physical and health capital; (iii) health status of worker; (iv) new knowledge. In addition, $\frac{1-\omega}{\omega}$ can be shown as the speed of knowledge diffusion. The problem of equation 5.11 in estimating the link between health capital and income at individual level is that it can reverse causality, because output growth may influence the rate of input growth. To overcome this problem, Bloom et al. (2004) suggest that 'we use lagged levels and growth rates of inputs and output as instruments for current input growth rates' (Bloom et al., 2004, p. 6). In equation 5.11, the coefficients of all variables contain the effect of vintage knowledge. Therefore, we can estimate the importance of knowledge (i.e., technical progress) to the other variables related to the process of production in agriculture. However, with a two stages regression model, variables that remain constant with time such as family background, culture, and socio-economic indicators will be omitted. To catch the effects of those variable types, we can regress separately equation 5.3 in each year; the purpose of equation 5.16 is to get $\frac{1-\omega}{\omega}$, and then we can calculate ω.

In the literature, it is suggested that in developing countries where low income is a major characteristic, food expenditure is an important proxy for health status. This means that the proportion of food expenditure has the most significant effect on health status, while health expenditure is smaller and individuals only spend their money on health facilities when they are ill. Diseases are accidental and they are uncertain. When they happen, households face two options: to cure or not to cure. When they choose to cure, they have to reduce their food expenditure; when they choose not to cure, food expenditure can remain unchanged. Sometimes, households have no choice but to cure, such as when a member has caught a serious disease. They will then spend money on health whether their budget can afford it or not.

The comparison between parameters inside and outside parentheses in the last equation is used as a standard test for the substantial correction of the model. Failure to reach the test implies that we need further consideration on the quality of data and the number of observations.

In order to estimate the determinant of health investment on households' income, the following equation has been estimated:

$$y_{i,t} = a_{i,t} + \alpha k_{i,t} + \beta health_{i,t} + \theta_1 s_{i,t} + \theta_2 exp_{i,t} + \theta_3 exp_{i,t}^2 + \theta_4 h_{i,t} + \theta_m z_{it}' + \varepsilon_{i,t}$$

To estimate the above equation, at first, OLS regression will be done to gain baseline information; then, GLS random and fixed effects will be estimated in order to glean in depth information for determining an appropriate estimation. Furthermore, an instrumental variable regression strategy will be considered as the final estimation, because IV regression is used in the case of omitting variables, or Ordinal Least Square Regression has an endogeneity problem. However, IV regression is only employed when instrumental variables are available for selection. In my thesis, IV regression is used because we do not have variables that indicate health status. In health economics literature, health status has positive correlation with income. This means that a person with good health will have a higher income. Thus, I used income in the last period as a proxy or health status. In fact, income in the last period has caused from many factors such as education or experience. I therefore employed residual in the equation between incomes in the last period with other things to proxy for health status in the past. Additionally, investment has a lagged effect on income; therefore, investment in health in the past will affect results in the present and future.

5.5. Data and variables statistics

5.5.1. Datasets description

The data used in this chapter was derived from the Vietnamese Household Living Standard Survey (VHLSS) 2002 and 2004.

The VLSS 1993 was conducted by the State Planning Committee and General Statistics Office with financial aid from the United Nations Development Program (UNDP) and the Swedish International Development Agency (SIDA), and technical support from the World Bank (WB). VLSS 1998 was carried out by the General Statistics Office with financial contributions from UNDP and SIDA, and technical assistance from WB. VLSS 1993 had 4,800 households while VLSS 1998 contained 6,000 households. VLSS 1993 and VLSS 1998 covered 61 provinces in Vietnam.

VHLSS 2002 was conducted by the General Statistics Office with financial support from the Japanese Bank for International Cooperation and technical assistance from WB. The VHLSS had two versions: the first version has a small questionnaire (36 pages) with the sample covering 60,000 households; the second version had a larger questionnaire (43 pages) with a smaller sample of about 15,000 households. The larger questionnaire is comparable with the VLSS questionnaire; however, the VHLSS 2002 questionnaire included some parts, such as anthropometrics, migration, and savings and credit, which were not present in the VLSS. The differences between VLSS 93, 98 and VHLSS 2002, 2004, and 2006 can be seen in the following table.

Table 5.1. Summarises the different characteristics across the three household surveys

Name	Period of data collection	Sample size	Length of the household questionnaire	Lowest level of representativeness	Types of data collected
1993 Vietnam Living Standards Survey	1992–1993	4800	110 pages	Seven regions	Household member characteristics, education, health, employment, migration, housing, fertility, agriculture, non-farm self-employment, expenditure, assets, other income, and savings and credit.
1998 Vietnam Living Standards Survey	1997–1998	6000	110 pages	Ten strata (7 rural regions and 3 types of urban areas)	Almost identical content and structure as the 1992-93 VLSS.

Name	Period of data collection	Sample size	Length of the household questionnaire	Lowest level of representativeness	Types of data collected
2002 Vietnam Household Living Standards Survey	2002	15,000	43 pages	Urban and rural areas of eight regions	Similar to the VLSS but no migration, anthropometrics, savings, or credit modules. Other modules simplified.
		75,000	36 pages	61 provinces	Similar to small-sample VHLSS but no expenditure module.

The VHLSS 2004 and 2006 questionnaires were almost identical to the VHLSS 2002 questionnaire. In VHLSS 2004 and 2006, the part related to households' health status was expanded. Additionally, some variables are present in the VHLSS 2004 and 2006, but not in the VHLSS 2002. This problem may lead to biased conclusions. To overcome this issue, we should avoid omitting variables as much as possible.

5.5.2. Variables Statistics Summary[28]

In order to have an appropriate estimation, agricultural households were selected from the dataset based on filtered indicators provided in the dataset. Agricultural households are households whose incomes are derived mainly from agricultural activities. Therefore, all households who earned mainly from non-agricultural activities were excluded from the sub-sample. Agricultural households' incomes, similar to other types of households, come from two sources, which are summarised in Table

---

[28] See appendix A for detailed information.

5.2. In fact, almost all agricultural households' incomes come from the first component, while the second component only appears in minority households. The small amount of agricultural households who have two components of income therefore do not have a significant effect on the empirical results. Income of agricultural households was measured in Vietnamese currency; and has been taken logarithm in doing empirical task in order to show the effect of human health capital on the rate of income growth.

Table 5.2. Income components

Total annual agricultural household income	First component of agricultural household income	Earning from agricultural activities in the last 12 months
		Earning from non-agricultural activities in the last 12 months
	Second component of agricultural household income	Overseas and domestic remittance from people who are not household members
		Pension, one-time sickness, and job loss allowance
		Social welfare allowance
		Income from insurance
		Interest on savings and the like
		Income from leasing
		Income from charity organisations, associations, and firms

Source: Summary from questionnaire of VHLSS in 2002, 2004, and 2006.

For the land variable, which proxies households' fixed capital, I adopted a three stages method as follows. *Firstly*, only households that use land for doing agricultural activities were extracted from these surveys. Therefore, all households had land, but those not partaking in agricultural activities were excluded. The purpose of this step was to focus on households closely and directly related to agricultural activities; therefore, the subsample will provide more accurate results. *Secondly,* agricultural land is the total of land area and water surface where agricultural households work and earn incomes from these activities; consequently, the land variables cover planting, aquaculture, and forestry land. In this instance, most households did not have all types of land, and their land areas were the total of one or two types of land. This variable also reflects the types of activities that households are doing for earning an income. *Thirdly,* taking the natural logarithm of household land was the last step; land was measured in 1,000 meter squared. In terms of economic principles, it is hoped that this variable will have a positive effect on income. Sign of this variable reveals the positive correlation between income and fixed assets in the economic literature.

Food expenditure was another important variable; it was measured in Vietnamese currency and taken in logarithm form in order to implement estimation. Food expenditure is all expenses of a particular household on food. In the case of agricultural households, food expenditure includes two parts: (i) spending on food; (ii) other sources of food in the households such as vegetables, chicken, and pig. In this chapter, food expenditure was selected as a proxy for nutrition intake.[29]

---

[29] It is understood that food expenditure is not a perfect proxy for nutritional intake. It is better to have a qualitative aspect, such as individual taste, as a proxy for nutritional intake. However, it is hard to find a suitable dataset that can fulfill the demand of my study. In fact, there is no dataset in Vietnam that can provide information about both income and nutritional intake. I have already considered three datasets: the Vietnamese Household Living Standard Survey, the Health and Demographic Survey, and the Population Census. Only the first dataset provided information about household income and

Health expenditure includes all expenses of a household on health. Health expenditure, thus, has three parts: (i) expenses on regular health checks; (ii) expenses on emergency cases; (iii) expenses on medicines and medical equipment (if applicable). Health expenditure was measured in Vietnamese currency and in logarithm form in order to see the effect of health expenditure on agricultural households' income fluctuations.

Education expenditure was measured in Vietnamese currency. It included all household spending on education such as tuition fees, volunteered contribution to children's schools, and other fees for extra or further education of a household member that might affect their income in the future. Education expenditure has been taken logarithm to make estimation. Additionally, education attainment was measured according to years of schooling.

Life expectancy was calculated from Vietnam's population census data and measured in years. However, the difference between the VHLSS and population census in terms of sample did not allow us to have life expectancy data at an individual level. Therefore, life expectancy was used here as aggregated data at the provincial level. The timing of the population census and the VHLSS differed, so life expectancy in this instance was used to proxy the general health conditions of province j in the time t.

The effect of other household variables such as family background, geography, and gender of the head of a household were captured by variable $z'_{i,t} \cdot z'_{i,t}$, and is therefore a set of variables that includes characteristics

---

food expenditure or consumption. Furthermore, because my thesis concerns investment, I have selected variables that can be measured in currency units. Thus, expenditure on food was selected in the lacking database circumstances as it concerns Vietnam.

of households at the individual level and some dimensions where gender[30] or ethnicity remained unchanged over time. Family background was measured by the number of family members. The gender of the head of a household was a dummy variable where the value of 1 was gained where the head was male and a value of 0 for female. The geographic variable was another dummy variable with two values, 1 and 0, to indicate the region where the family lived.

5.6. Estimations and results

To estimate the effects of individuals' backgrounds and their human capital investment on their income, we first estimated equation 5.3.[31] The results are presented in Table 5.3 and 5.4 (see appendix B). The results in these tables include physical capital (represented by land area), human capital (characterised by food and health expenditure, years of schooling, experience and experience squared—as suggested in microeconomics— and life expectancy at birth). Other effects are gender, family background (symbolised by the size of households), and geographical variables, which are variable indicators.

Per our expectations, land and food consumption had a positive effect on income. The effect of land on agricultural households' income in 2002 was higher than in 2004, reflecting that the source of households' income was changing patterns. This result implies that farmers are no longer

---

[30] In Vietnam, changing of gender is abnormal, so it is assumed that gender remained unchangeable over time.

[31] For detailed information, see appendix B.

relying only on their land. Besides working in the fields, they are also participating in other economic activities in the city near their towns. This change (from 0.73 in 2002 to 0.08 in 2004) is quite large and reveals the fact that farmers are leaving their homelands to find better ways of earning more money. This situation will affect their pattern of nutrition intake, which will in turn influence their income (represented by their human health accumulation). Consequently, the effect of food expenditure on agricultural households' income should be raised, a notion that the regression results support. It was found that the coefficients moved from 0.19 in 2002 to 0.43 in 2004. This implies that nutrition intake has an important role in household income. Nutrition intake relates not only to food expenditure but also to the consumer behaviour of households in terms of food. Higher income households tend to spend more money on more nutritious food and better meals than lower income families. As a result, richer agricultural households have better health than other households. The decreased effects of land and the increased influence of food expenditure on household income fit microeconomics and agricultural economics, because when income is increased, farmers will want to leave the agricultural sector; to do so, they have to prepare their health (ability to gain high productivity) by adjusting their nutrition intake.

Health expenditure of agricultural households tells us the different aspect of their behaviour. It is assumed that health care consumption will take place when farmers have diseases or bad illnesses. In developing countries like Vietnam, visits to the doctor or regular health checks are uncommon, especially for agricultural households. Therefore, health expenditure is the price of recovering the ability to work. In this sense, the coefficient of health expenditure represents the part of income that may be lost if farmers do spend money on a cure. The coefficient of health expenditure variable increased from 0.02 in 2002 to 0.06 in 2004. When households did not pay for using health care, the lost part of income was around 0.02 percent in 2002 and 0.06 percent in 2004. The increasing of this coefficient emphasises the vital role of health in terms of income. The approximately fourfold raising in the effect of health expenditure also means that farmers are more aware of their health status.

The goal of increasing income and wealth has led farmers to better consider their children's future. Investment in education offers their children a better life in the future. This expectation is a current dynamic of the hardworking life of agricultural households, where incomes remain lower than in other sectors. Therefore, the educational coefficient expresses the push effect of earning more income, which increased from 2002 to 2004. In 2002, the coefficient value was about 0.06 percent, while this value was around 0.08 percent in 2004. The difference between the two coefficients reflects the increase of parents' expectations for their children's futures. In addition, increasing expectations should be considered an important dynamic of investing in the health of farmers, because they understand that with good health they can earn more money with which to save for their offspring's future. This way of thinking is common not only amongst agricultural households, but also amongst industrial households.

Another human capital indicator is years of schooling. The effect of years of schooling on household income is generally positive and smaller than other studies in the literature of economics of education (Barro, 1999; Barro & Lee, 2001; Psacharopoulos, 1994; D. T. Tran, 2009; D. T. Tran & Do, 2007). The coefficient was 0.01 in 2002 and 0.04 in 2004. This means that one more year of schooling increased 1.3 percent and 3.75 percent of income in 2002 and 2004, respectively. Both coefficients are statistically significant at 1 percent level and consistent with some types of regression. This result shows us the consistent relationship between income and years of schooling, as suggested in microeconomics and labour economics (for more information, see chapter 6). In the literature of development economics, the rate of return to schooling is 9.1 percent (Psacharopoulos, 1994); in this instance, however, the rates in 2002 and 2004 were 6.72 percent and 8.6 percent, respectively. Nonetheless, we cannot reject the hypothesis that the rates in 2002 and 2004 were equal to 9.1 percent (F statistic is 1.6 and 0.18 for 2002 and 2004, respectively).

The coefficients of experience and experience squared in 2002 are not consistent with microeconomics due to their sign, but the test of joint zero provides strong evidence that we cannot reject the hypothesis that both of them are zero. Conversely, in 2004, despite the fact that the coefficients of them are not statistically significant, their signs are suitable to microeconomics, and then the hypothesis that both of them is zero can be rejected at 10 percent level. The reason for the poorly determined coefficients is that experience and experience squared are highly correlated (0.97). The poor result may therefore be due to the wide variations in the experience variable.

Life expectancy is an important proxy of human health capital. This variable usually has a positive effect on income. Both regression results using 2002 and 2004 cross-section and panel data gave us the strong sense that life expectancy has a positive effect on agricultural households' incomes. However, life expectancy was not well determined in the 2004 dataset, while it was statistically significant in the 2002 dataset.

There is consensus about this variable, and many researchers agree that life expectancy should be seen as a valuable variable for measuring human health capital. The contradictory results in this instance led us to using instrument variables (IV) for life expectancy. Life expectancy is a complex and multidimensional variable. In the literature, life expectancy depends on the level of income and the natural environment people live in. However, we cannot use income here as an instrument, because income is used as a dependent variable. Should we as a result use education expenditure instead of income? The answer is yes, because in the economic literature, high income households tend to have high education expenditure.

Other indicators concern the social and natural environment; household size was used as a proxy for family background and regional variables represented natural conditions. The result of instrument variable regression is presented in Tables 5.3 and 5.4 in appendix B. The results are better than before in terms of determining the effect of life expectancy on agricultural household incomes. The effect of life expectancy on income was 0.06 in 2002 and 0.03 in 2004. This means that increasing life expectancy by one year enhanced farmers' productivity and agricultural output by about 6.5 percent in 2002 and 3.4 percent in 2004.

The next step is to test the instrument variable regression model. We applied Hausman's test to check whether IV regression was suitable or not. The result showed that IV regression was a better fit than ordinary least square (OLS) regression. The $\chi^2(8)=35.59$ implies that OLS regression was inconsistent here and that IV regression should be applied. The second test was an endogenous test for life expectancy. Thus, we had to test the hypothesis that life expectancy is an endogenous variable. H_o hypothesis is: Life expectancy is an exogenous variable and H_1: Life expectancy is an endogenous variable. Failure to reject the H_o hypothesis will provide strong evidence that IV regression is inconsistent. We used two tests: the Durbin score and the Wu-Hausman test. The results were very good, and saw the H_o hypothesis soundly rejected at 1 percent level. The Durbin score was 24.94 and the F-test of Wu-Hausman was 25.09. The conclusion in this instance was that IV regression presents consistent results.

Family background (represented by household size) has complex effects on the income of agricultural households. In 2002, family background had a negative effect, while in 2004 it had a positive effect. The changing patterns in family background may explain this result. We cannot however conclude here that family background is responsible for these effects on household income, as it is too soon for reaching such a conclusion, given the lack of information. Instead, we emphasise that family background has an effect on income (because in all regression results household size was statistically significant at 1 percent level).

The effect of gender on income is captured by the coefficient of female variables. The gap between male and female in terms of income is quite large; however, this gap is becoming smaller. Female-headed households had a 17.6 percent lower income than those headed by males in 2002. In 2004, this gap was only 4.8 percent. The change is significant and reflects the situation that government policy is efficient in trying to close the gender gap in rural and agricultural areas. Another reason for the change stated above is that women in agricultural and rural areas are now more dynamic and energetic than in the past in their approach to finding opportunities for earning money. They have better health and encouragement for doing business other than agricultural tasks. Male and female farmers find it easier to improve their health and their chances for joining the labour market and find it less problematic to travel from their homeland to the city to trade their output. In conclusion, women's health capitals are better than before, which is reflected in the narrowing gap between males and females in terms of income.

To estimate the effect of regions on household income, some indicator variables were added to the model. Interestingly, the regression results suggested that regional variables did not have a contradictory effect. In 2002, most regions had a negative effect on families' income, while in 2004, all regions had a negative influence on households' income. Thus, households in all regions did not have any advantage in their opportunities for gaining higher incomes. In other words, all households had the same opportunity for gaining income. This result was an achievement on the part of the Vietnamese government in levelling the income gap between regions. However, the important implication in this instance is that if farmers did not have any assets and/or chances to improve their health, their incomes would have been decreased no matter where they lived.

In addition, equation 5.3 is regressed by eight regions and the results are reported in Tables 5.6 and 5.7 (see appendix B). Overall, the results for all regions were similar in terms of microeconomics, where most signs of coefficients were substantial in 2002 and 2004. The decision to omit educational expenditure made our model better in terms of microeconomics. The signs of experience and experience squared were considerable in both regressions, while the sign of household size changed from negative in 2002 to positive in 2004, which reflects the complex effect of family background on families' incomes. The rate of return to schooling in 2002 was around 10.1 percent and in 2004 around 9.3 percent. These rates are close to those in the literature concerning the microeconomics of education. Additionally, the effects of land and food expenditure were positive in all regions and for each region, and absolute values varied by region, while health expenditure reflected a more complicated picture. In 2002, the coefficients of health expenditure were reported differently in terms of sign of coefficient, while in 2004 they were positive for every region. Moreover, life expectancy did not have a unique sign of effect on household income in terms of regional differences in 2002 and 2004. These results suggest that the effect of life expectancy on income had been influenced by geographical conditions. Thus, natural conditions have a significant effect on life expectancy and on the health capital of individuals.

In the context of personal characteristics, represented by the gender dummy variable and family background as presented by household size, the situation is similar to the above cases. Female-headed households had lower incomes than those headed by men, except in the Red River Delta, North West, South Central Coast, and South East in 2002, and the Red River Delta, North West, North Central Coast, and Mekong River Delta in 2004. Almost all of these coefficients were statistically insignificant. In these regions, women had higher incomes than men.

We now begin to estimate equation 5.11 under the assumption that fundamental innovations were constant and provided by the government. The results are reported in Table 5.5 (see appendix B). The parameters of the regressions were estimated by instrument variables two stage least squared, where all lagged inputs were instrumented.

The results in column 1 of Table 5.5 include only physical capital, nutrition intake, health expenditure, and life expectancy. Our results suggest that increasing land by 1 percent would increase income by approximately 25 percent, while enlarging nutrition intake by 1 percent would raise income by nearly 23 percent. Health expenditure had a far smaller effect; the coefficient was 0.02 implying that spending money on healthcare 1 percent more would gain a 0.02 percent increase in income. Life expectancy in this case was statistically insignificant, but the coefficient was similar in the estimation of Bloom et al. (2004). The positive coefficient means that a one year increase in life expectancy will contribute to an increase of about 4 percent in income. However, we cannot reject the hypothesis that this coefficient is equal to zero. The coefficient of speed of knowledge diffusion was 0.16, implying that a 1 percent increase in spreading knowledge would lead to a nearly 0.16 percent increase in income. Additionally, the coefficient led to the calculating of the fraction of vintage knowledge in equation 7. In this situation, $\omega = 0.86$; therefore, just over 86 percent of income is made by experienced knowledge.

Adding the schooling variable in column 2 has the effect of rendering all human capital variables significant, where only growth of health expenditure is significant at 10 percent level. An increase of 1 percent in agricultural land area led to about a 0.26 percent increase in households' income, while raising nutrition intake by 1 percent brought about a nearly 0.2 percent rise in families' earnings. The effect of growth in health expenditure to growth of income was slightly smaller than the first regression, while the effect of life expectancy increased considerably. One year more of schooling had an increase of about 0.04 percent in households' income. The coefficient of life expectancy implied that a one year increase would add up to 0.17 percent to households' income in the period 2002–2004. However, the mean of life expectancy growth in this period was only 0.48 years. Thus, life expectancy actually had an effect of about 8 percent on agricultural households' profits. This coefficient was significantly stronger than the other estimations in the literature. In the second regression, the coefficient of the speed of knowledge diffusion was 0.12, smaller than the first regression. Consequently, the coefficient of vintage knowledge $\omega = 0.89$ was calculated using equation 5.3.

Adding experience and experience squared variables to the estimated equation we now have the full model in column 4. As can be seen in the table, growth of land was a significant source of agricultural households' income in the period 2002–2004, while the determinant of nutrition intake was slightly smaller (0.21 compared to 0.26). As such, the coefficient of health expenditure showed us that its marginal effect to income was around 2 percent, whereas one year more of schooling could increase household heads' income by nearly 4 percent per year. Experience and experience squared are still not statistically significant, due to the reason mentioned above. The effect of life expectancy on income was just over 0.16 percent, meaning a one year increase in life expectancy would raise income by 0.08 percent per year. However, the coefficient of the speed of knowledge diffusion was the smallest in four estimations, only 0.11. This led to $\omega = 0.90$.

In conclusion, while our estimated results were consistent with microeconomics and our model well defined, some variables such as experience and experience squared were not well calculated. However, most other human health capital variables were statistically significant and had a strong sense of economic meaning. The determining of them based on income was considerable and showed that evidence of human health capital investment at the household level in the Vietnamese agricultural sector was similar to other results in the literature of economic development studies, microeconomics, and agricultural economics worldwide.

5.7. Policy implications

Vietnamese farmers highly value their health as long as their income is lower than their demand. Consequently, nutrition intake plays an important role in maintaining their ability to work and enhance their productivity. In the microeconomic literature, it is suggested that one more year of schooling raises earnings by roughly 10 percent and approximately 0.6 years in terms of life expectancy. In this study, it was found that one more year of education increased life expectancy by 0.29 and 0.33 years in 2002 and 2004, respectively (with the assumption that the discounting rate is nil). In a panel 2002–2004 dataset with smaller observations, we found the rate to be only 0.06 years. This result implies that in agricultural situations, the rate of return of education to life expectancy in one period was far higher than in the two periods of time. Therefore, public policy should integrate this point in the decision-making process. Development achievements, at the microeconomic scale, are expressed as the improvement of household living standards. In the case of Vietnam, expenditure on food was the most important indicator for explaining the development process, especially for agricultural households. Based on the above estimation, the question raised here is what policy government should implement to improve development attainments.

Firstly, nutrition intake is an important input for increasing agricultural households' income. Nutrition intake plays a vital role for people, but in the context of the Vietnamese agricultural sector, food expenditure has a more significant effect on income. This implies that government should maintain the availability of access to food markets and provide fresh food products to households. Building markets close to villages, and making it convenient to access them, should be government priorities in any rural development strategy. In 2004, 62.1 percent villages had communal or inter-communal markets, and the proportion of hamlets having daily, periodical, and whole sale markets was 28 percent, 10.7 percent, and 4.6 percent, respectively. Additionally, the average distance from hamlets to the nearest daily, periodical and whole sale markets were 4.9, 2.7, and 6.8 kilometres, respectively. The percentage of hamlets that had markets was low; thus, building more markets at the hamlet level is a first priority. Building markets at the provincial level is the second priority, because the percentage of the population that has markets at hamlet level is smaller than those who have access at provincial level. In the condition of a limited budget, local authorities should focus on the micro-scale instead of the macro-scale. The reversal of policy targets from top-down to bottom-up styles will improve the development process of Vietnamese agriculture and the Vietnamese economy as a whole.[32]

The second approach is that local and central governments have to control and examine the quality of foodstuffs supply by their science laboratories. Researchers who work for state institutions and laboratories have a responsibility to investigate bad food supply sources to protect the human health capital stock embodied in farmers. Coordination between researchers and policymakers is necessary to provide safe conditions for a fresh food supply. With better availability of fresh foodstuffs, households have a better chance of arranging healthy meals. Accordingly, the health of household members can be improved. If government can process both

---

[32] For a more detailed analysis, see chapter 7.

of the above approaches, human health capital stock will be maintained and improved.

Secondly, estimation results suggest that one more year of schooling will lead to bigger incomes; this implies that policymakers should focus on reforming the education system at the local level as an approach for enhancing the educational attainment of agricultural households. People with higher educational attainments, such as bachelor's degrees or higher, will leave the agricultural sector to find better jobs. Thus, authorities should use farmers who have obtained secondary certification at the highest educational level as a target group of such a policy.[33] Vietnamese farmers usually start work in their fields after finishing essential education; they therefore work based on their experience and their parents' experience, which is expressed by the high value of ω above. This situation can only exist in the short term, however. In the long term, working by experience cannot bring agriculture into a sustainable way of development with high productivity.

Lessons from many other countries in the present and in the past concerning developed experience have suggested that the Vietnamese government has to focus on training their current farmers. The training aims to provide farmers with adequate knowledge in applying the newest achievements in fundamental innovations related to agriculture. Therefore, the following solution is suggested: reform the educational system at the provincial level, encouraging secondary students to study at vocational schools with main subjects related to agriculture. The aim of the training

---

[33] In our empirical framework, farmers with an educational level higher than secondary level are excluded for two main reasons. Firstly, these individuals are old; secondly, they do not represent common trends in agriculture.

is so that farmers can apply new technology with limited instruction and enable them to give researchers feedback.

Thirdly, young people with better health and educational attainment always move to cities where they are able to find better employment, while the elderly, women, and those with lower educational attainment tend to remain in agriculture. These people need better healthcare, and they consider health expenditure an investment, because their incomes are low. Healthcare fees are high and farmers cannot get health insurance if the government does not subsidise it or introduce other supportive policies. Lacking healthcare will deteriorate farmers' health capital stock. Consequently, agricultural households' productivity may be lost.

To overcome this situation, the Vietnamese government issued a policy that allows the government to spend a part of the state budget to subsidise health insurance fees for farmers. Under the new policy, farmers are paid a full or partial healthcare fee, depending on their type of disease.[34] However, there are numerous problems with this policy due to the administrative process. For example, processing time is long and progress is slow; therefore, many patients do not use their insurance cards. Secondly, the subsidy is lower than the demand of farmers. Based on the existing system, the Vietnamese government should simplify the administrative process, with stronger contracts between the government, insurance companies,

---

[34] For common and non-serious health problems, agricultural households have been paid full healthcare fees, while serious health problems are paid for partly by the Vietnamese government. There are nearly 14.3 million Vietnamese people (approximately 17.4% of the total population), supported by the government with annual funds of 1.300.000 VND (nearly 60 AUD/year) via local government budgets set aside for paying healthcare fees.

and awardees. Additionally, the government should increase funds for subsidising health insurance aimed at agricultural households, where incomes are lower than in other sectors.

In conclusion, agriculture is a vital sector of the Vietnamese economy and needs better attention from the government in terms of human health capital investment. Agricultural households play an important role as consumers, while the government acts not only as the largest supplier, but also as the largest supporter of farmers. Under the mechanism called 'market-oriented socialist economy under state guidance', the Vietnamese government is the second most important player for enhancing agricultural productivity and agricultural households' incomes. Maintaining the relationship between agricultural households, state, researchers, and firms, and a supportive policy in terms of improving living standards are two ways of boosting agricultural development in Vietnam.

5.8. Conclusions

In this chapter, I presented results from different regression strategies, as I wished to illustrate the varying coefficient values of different regressions. In fact, the coefficient values differ only slightly, their standard errors different due to varying regression strategies. However, in all regression strategies the core variables, for instance health expenditure, had positive signs. This means that investment in health has an investment effect rather than a consumption effect. In the introductory chapter, I stated that there is no clear border between consumption and investment. If an increase in consumption leads to an increase or decrease in income, as represented by positive or negative coefficient values, we can conclude that consumption has an investment effect, where a positive sign shows good investment and a negative sign shows bad investment. On the other hand, if an increase in consumption does not lead to an increase or decrease in income and the coefficient value is equal or close to zero, we will conclude that consumption has a consumption effect. In order to know the effect of consumption on income, we need to employ estimated results for conclusions to be made. Therefore, in this chapter, I used various econometric methods to establish whether health expenditure in the case of Vietnamese farmers has a consumption or investment effect.

In this chapter, IV regression was used, because omitting variables or Ordinal Least Square Regression presents an endogeneity problem. However, IV regression is only employed when instrumental variables are available for selection. Essentially, IV regression was used because we did not have any variables indicating health status. In health economics literature, health status has a positive correlation with income (J. Strauss & D Thomas, 1998). This means that a person with good health will have a higher income. I therefore used income in the last period as a proxy for health status. However, income during the last period resulted from several factors such as education and/or experience; as a result, I employed residual in the equation between incomes during the last period, with other variables to proxy for health status in the past. Additionally, investment has had a lagged effect on income; thus, investment in health in the past will have result in the present and future income.

Additionally, I also presented my estimates of the effect of investing in health on the income of Vietnamese agricultural households for 2002–2004. Based on the theory of endogenous growth, the results support the idea that learning by doing has significantly impacted household incomes. In the context of health, it was found that around 85 percent of the method of working at the present time is influenced by past experience. This situation is due to people's health status. Moreover, nutrition intake represented by food expenditure corresponds significantly to income. This means that in the case of Vietnam, food expenditure is a vital source of income and a good proxy for nutrition intake. Health expenditure, however, has a less significant effect on household incomes than nutrition intake. As such, improving life expectancy may increase household incomes by over 10%.

Finally, there are some limitations to the study due to the availability of data. I did not have access to a dataset with the complete information needed for my study, and believe that my results would have been better if I had had more data. My estimated model might be true for this dataset, but might not work well with other datasets. Thus, in future research, I will consider other datasets in Vietnam or other countries that have similar development characteristics to Vietnam in order to test the empirical results of my model.

Chapter 6

EDUCATIONAL INVESTMENT AND AGRICULTURAL DEVELOPMENT IN VIETNAM

6.1. Introduction

This chapter provides estimates of the private rate of return of investment in education of agricultural household members to agricultural household income, based on three waves of observations on a sample of farmers of the Vietnam Household Living Standard Survey in 2002, 2004, and 2006. The starting point of the analysis is basic human capital investment theory, which is taken seriously as a potential vehicle for estimating the rate of return of educational investment of agricultural household members on their income. Moreover, the theory explains observed patterns of school attendance, work and income. The Hausman and Taylor (1981) (hereafter HT) estimation framework that is adopted fully permits an investigation of whether individual unobservable effects can be estimated. It also works well in fitting observed data to estimate rate of return of investment in education regarding school attendance and work and income in the VHLSS farmers' data. Additionally, a reasonable forecast of future work decisions and income patterns could be produced.

The HT approach provides accurate interpretations for the parameters that are estimated, which has several important consequences. *Firstly*, parameters that are estimated may be of interest in their own right, for instance, technology. In the context of the thesis, my framework separates the quantitative significance of school attendance and specialised experience in the production of occupation-specialised skills. *Secondly*, because the thesis focuses only on agricultural households, it is able to quantify the specific effect of learning decisions on agricultural household incomes. In the current case, the monetary incentive for attending college and attaining a higher education does not receive adequate attention, because farmers in Vietnam do not wish to study at higher levels. Farmers would rather learn at vocational training centres than study at university. Previous research on this topic in Vietnam has generally treated school and working decisions in isolation and has therefore been limited in its ability to address this question. The HT approach, conversely, could provide a suitable method that allows for treating school attendance as an endogenous variable and thereby estimate the individual unobservable effect. *Finally*, the HT approach allows for employ time-variant variables and time-invariant variables in estimation, and removes the correlation between error term and included variables that are usually neglected by random-effects estimation.

In order to understand the contributions of this chapter, it is useful to regard it within the context of current human capital investment literature. The general theory of human capital accumulation originated from the interpretation of life cycle earnings profiles. From its foundation (Mincer,

1958; Becker, 1964; Ben-Porath, 1967), the theory has been developed relative to observable measures of human capital investments, which is usually school attainment. Consistent with an investment framework, the more recent economic literature has focused on estimating rates of return to schooling. Based on the comparison of age-earnings profiles between groups who graduated from schools, this early 'rates of return' referred to schooling as a signal for assigning individuals into the population.

These calculations ignored the fact that school attendance (or human capital accumulation) is a choice (Keane & Wolpin, 1997) and took the view that if individuals had the same endowment and faced identical financial constraints, they would behave identically with respect to the choice of schooling (Rosen, 1977). However, individuals are not identical at all and have their own characteristics. Therefore, the rate of return estimation should address the effect of individual differences in schooling decisions.

This chapter extends earlier work by considering the individual unobservable effects in estimating the rate of return to schooling in the case of Vietnam. This chapter, therefore, applied the HT approach for estimating the rate of return to schooling of farmers in Vietnam from 2002 to 2006. Despite the short time period, the empirical findings have important implications on the education policies of the Vietnamese government. Additionally, empirical findings show the different rates of return to educational investment across regions of Vietnam. This implies that the implementation of polices might consider the specification among regions.

This chapter is organised as follows: in section 2, general information about the education system of Vietnam will be presented. Section 3 will present the structure of the human capital investment model and discuss the method employed. Section 4 describes the general statistical summary of the data and variables that will be used in estimation. Section 5 presents empirical findings and discusses the implications of the model. The implications of the model may be categorised into two important areas: (i) the importance of individual unobservable effects in estimating rate of return to schooling; (ii) the important implications on the development of vocational training for farmers. The last section presents conclusions and further study.

6.2. A brief review of education in Vietnam

Education, higher learning systems and vocational training schools are very important to business, as they provide good quality labourers. They also supply a system for transferring new knowledge and skills and to train new employees for the modern business environment.

Vietnam has experienced dramatic changes since the 1990s. The economy experienced a boom from 1990 to 1996. A downturn occurred in 1997–1998 due to the financial crisis in South-East Asia, followed by a recovery. Between 2000 and 2007, Vietnam continued to slowly develop. The average rate of growth from 1991 to 2008 was 7.6 percent annually. The success of the Vietnamese economy has been documented by many researchers worldwide (Athukorala, 2009; Leung, 2009; Menon, 2009; Ohno, 2009; Pincus, 2009; T. Q. Tran, 2009; Vo & Nguyen, 2009). *Doi moi* (renovation) policy is seen as a major reason for economic achievement in Vietnam and can be divided into two phases: (i) *Doi moi* I (1986–1996) and (ii) *Doi moi* II (2001–2007) (Leung, 2009). According to Leung, *Doi moi* I opened the national economy and connected Vietnam to the international economy, while *Doi moi* II concentrated on releasing the domestic private sector and deeper integration with the global economy, which coincided with Vietnam becoming the 150th member of the World Trade Organization in 2007. The future of economic growth in Vietnam relies heavily on a higher quality of workers who retain better and more modern knowledge and skills. Currently, a large percentage of the labour force working in agriculture needs to be retrained by vocational training schools to prepare them for working in the industry sector.

Education plays a vital role in Vietnam's culture and society. It is viewed as a path for advancement, where families make trade-offs to ensure that their offspring have access to adequate education. Parents generally believe that a better education leads to a better future for their children. As a result, parents spend much of their time, attention and energy on their children's education.

6.2.1. Education in Vietnam

The education system in Vietnam has four levels: (i) kindergarten; (ii) general education; (iii) vocational training; (iv) higher education. Kindergarten takes five years to complete. General education has three levels: primary, lower secondary and higher secondary. Vocational training school is at the same level as higher education. Completing general education takes twelve years: five years for primary school, four years for lower secondary school, and three years for higher secondary school. After graduating from higher secondary school, learners have to take an entrance examination to study at university. At this stage, students have to choose university or college. Studying at university takes six years for medical school, while for other subjects the duration is only four years. Studying at college takes three years for all subjects. Students will generally spend one to two years studying to obtain a postgraduate degree and four years at the doctorate level. In total, from kindergarten to doctorate level, learners have to spend at least twenty-four years studying.

The education system in Vietnam is presented in the following figure.

Figure 6.1: The education system in Vietnam

6.2.1. Enrolment

General trends

Vietnam has made impressive expansion in enrolment in basic education since the 1990s. Figure 6.2 shows us the total enrolment at primary and secondary levels from 1987 to 2010. Vietnam achieves a high enrolment rate, despite having a low level of income per capita.

Figure 6.2: Trends in School Enrolment, 1987–2010

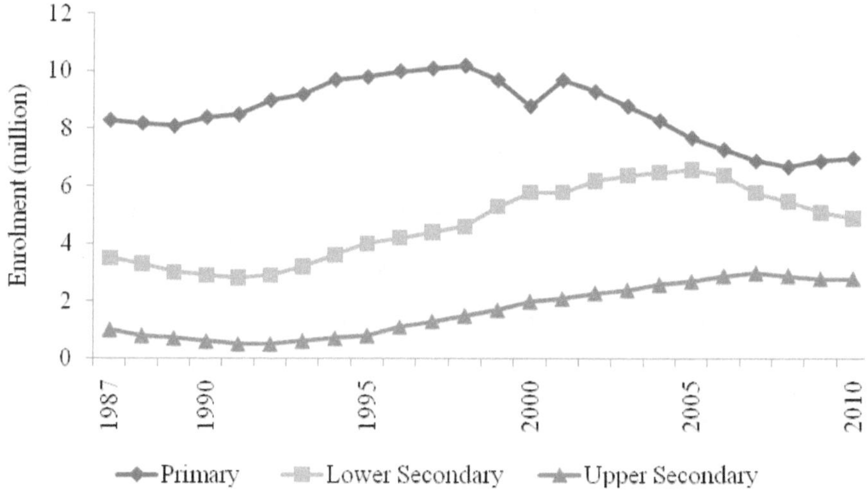

Source: GSO, various years

As can be seen in Figure 6.2, the overall trend of primary enrolment is at a decrease, while secondary enrolment for the same period has experienced an increase. The total enrolment in primary education decreased from 8.1 million pupils in 1987 to approximately 6.2 million pupils in 2010. On the other hand, the number of pupils who enrolled in lower secondary level increased from just below 4 million pupils in 1987 to 4.3 million pupils in

2010. At the same time, total higher secondary enrolment increased from 1.5 million pupils in 1987 to over 2 million pupils in 2010.

Gross and net enrolment

Education progress can be assessed by looking at the number of pupils who are studying at the school and the number of pupils who begin studying at the appropriate age. From 1993 to 2006, gross and net enrolment increased at all levels of education. Net enrolment rate of primary education increased slightly from 87 percent in 1993 to 89 percent in 2006. This rate of lower secondary education increased dramatically, from 30 percent in 1993 to approximately 79 percent in 2006. Furthermore, the rate of higher secondary education increased significantly from 7 percent in 1993 to roughly 54 percent in 2006.

Gross enrolment rate of primary education decreased moderately from 1993 to 2006, while this rate of secondary education increased considerably during the same period. From 1993 to 2006, the rate of primary education decreased by 15 percent, while the rates of lower and higher secondary education increased by 54 and 65 percent, respectively.

6.2.2. Education attainment by income group

Overall, the majority of Vietnam's population aged 15 and above has graduated lower secondary school. Statistically, 28.7 percent of Vietnam's population aged 15 and above graduated lower secondary school in 2006. Additionally, 24 percent of the population graduated primary school in 2006. Only 4.3 percent of the population had a vocational training certificate in 2006, and only 4.5 percent of the population had a higher education degree. On the other hand, 8.1 percent of the population has never gone to school and 14.5 percent do not have any certificate. For detailed information see Table 6.1.

Table 6.2 shows there is a difference between lowest income quintile and highest income quintile. In the first quintile, 27 percent of the population group has only a primary certificate, while in the fifth quintile, 17 percent of the population group graduated with only a primary education. For lower secondary education, nearly 27 percent of the first quintile group has a lower secondary certificate only, whereas just over 23 percent of the fifth quintile group finished at this level of education. For education levels higher than lower secondary education, percentages of the population who graduated only one of these education levels of income quintile 1 are overwhelmed by the percentage of the population in income quintile 5. Numerically, these percentages of the population of income quintile 1 are 6.5, 0.7, and 0.7 percent, compared to 19.0, 6.3, and 9.2 percent of the population in income quintile 5 for upper secondary, technical worker, and vocational education, respectively.

Table 6.1: Percentages of population aged 15 and above by highest certificate and income quintile in 2006

Unit: %

	Primary	Lower secondary	Upper secondary	Technical worker	Vocational
Whole country	**24.0**	**28.7**	**12.6**	**3.3**	**4.3**
Quintile 1	27.1	26.7	6.5	0.7	0.7
Quintile 2	27.3	32.8	9.2	1.5	1.5
Quintile 3	25.7	32.2	12.0	3.0	3.0
Quintile 4	23.9	28.7	14.9	4.6	6.0
Quintile 5	17.0	23.3	19.0	6.3	9.2

Source: GSO (2008)

6.3. Theoretical framework

6.3.1. The relationship between investment in education and economic growth

Measuring the relationship between investment in education and economic growth has attracted the attention of researchers worldwide. Academics have applied numerous distinguished economic theories such as the classical theory of economics, neo-classical theory of economics, and endogenous economic growth in this area of study. Among these theories, endogenous economic growth is the latest edition. It began with the extension of a typical Solow-Swan framework on economic growth by the Lucas-Romer endogenous growth model. The Lucas-Romer growth model suggests that endogenously accumulated human capital, presented by years of schooling, will directly affect labour productivity in a market economy. It is assumed that human capital is individual-specific; as such, innovation in the stock of human knowledge will be treated as an exogenous factor. Therefore,

$$Y_{i,t} = A_{i,t} K_{i,t}^{\alpha} (H_{i,t} L_{i,t})^{\beta}, 0 < \alpha, \beta < 1, \alpha + \beta = 1 \quad (6.1)$$

where Y is net income, K is physical capital, L is labour and H is human capital. Obviously, if the production function is a constant return to scale with physical and human capital, the model will be growth perpetual.

The above equation can be derived as follows:

$$Y_{i,t} = A_{i,t} K_{i,t}^{\alpha} L_{i,t}^{1-\alpha} H_{i,t}^{\beta} \quad (6.2)$$

Dividing both sides of the equation for $L_{i,t}$ we have

$$\frac{Y_{i,t}}{L_{i,t}} = A_{i,t} \left(\frac{K_{i,t}}{Li_{i,t}}\right)^{\alpha} H_{i,t}^{\beta} \quad (6.3)$$

Call $y_{i,t} = \frac{Y_{i,t}}{L_{i,t}}$ and $k_{i,t} = \frac{K_{i,t}}{L_{i,t}}$; therefore, $y_{i,t}$ is income per capita and $k_{i,t}$ is physical capital per capita. Taking a natural logarithm for both sides of the above equation we have

$$\ln(y_{i,t}) = \ln(A_{i,t}) + \alpha \ln(k_{i,t}) + \beta \ln(H_{i,t}) \quad (6.4)$$

It is generally if we assume $\beta \ln(H_{i,t}) = \rho_s E_{i,t} + \epsilon_{i,t}$, where ρ_s is private return to investment on education, $E_{i,t}$ is years of schooling and $\epsilon_{i,t}$ is error term, into equation (6.4), which then yields

$$\ln(y_{i,t}) = \ln(A_{i,t}) + \alpha \ln(k_{i,t}) + \rho_s E_{i,t} + \epsilon_{i,t} \quad (6.5)$$

We examined the determinant of investment in education of household members with a focus on heads of household and on household members' incomes. Thus, we needed to estimate the effect of human capital, presented by education attainment, on income at the individual level. The model above focuses only on observable effects, while unobservable effect is not included. The model is concerned with the effect of education investment on the income of the household; therefore, we need to consider the effect of household specifications such as cultural background, relationship among family members, the reputation of the family and other factors. These factors cannot be observed and are therefore referred to as individual unobservable effects. Therefore, it is advised that in addition to observable effects, there are many individual unobservable effects that affect household income.

6.3.2. Investment in education: Rates of return

In labour economics or economics of education, earning function is usually employed as a major factor for measuring rates of return of investment in education. Many studies have attempted to find a true theory for calculating rates exactly by using many different instrumental variables. Unfortunately, not all of these provide true results in estimating rates of return of investment in education.

The term 'internal rates of return of investment in education, measured by years of schooling' was initiated as a vital concept of human capital theory by Becker (1964). However, his method is not common practice in measuring rates of return of investment in education. Rather, the coefficient on schooling obtained from the log linear earnings equation is often viewed as a rate of return. Mincer (1974) popularised the model introduced by Becker and Chiswick (1966), which was later commonly referred to as Mincer's equation. The most popular formula of Mincer's equation is

$$\ln(Y_{x,s}) = \alpha + \rho_s s + \beta_0 x + \beta_1 x^2 + \varepsilon \quad (6.6)$$

where $Y_{x,s}$ is the earnings or wage at the age of x and the level s of schooling, ρ_s is 'rate of return to schooling'[35] and ε is the disturbance term with $E(\varepsilon|s,x) = 0$.

Let $Y(s)$ be a proxy for annual earnings of a farmer with s years of schooling, where s is supposedly constant over time. Let r be an external interest rate and T the length of a farmer's working life, which is assumed to not be reliable on s. The present value of income combined with years of schooling, according to Mincer (1958), is therefore expressed in the following equation:

$$V(s) = Y(s) \int_s^T e^{-rt} dt = \frac{Y(s)}{r}\left(e^{-rs} - e^{-rT}\right) \quad (6.7)$$

After that, equating earnings flow and taking log for both sides yields

$$\ln Y(s) = \ln Y(0) + rs + \ln\left(\frac{1-e^{-rT}}{1-e^{-r(T-s)}}\right) \quad (6.7)$$

The final term in the right-hand side will vanish when T is very large. This term is an alteration for predetermined life.

Model (6.8) implies that the more education people have, the higher income they will earn. When T is large enough, ρ_s in equation (6.6) will be close to r (interest rate). Clearly, internal rate of return of investment in education proxies for the discount rate of lifetime earnings at different education choices; in this model, it should equal to the interest rate. Consequently, ρ_s is an estimation of the internal rate of return. We therefore have three cases:

---

[35] The rate is assumed the same for all levels of education.

i. $\rho_s = r$, education market reaches equilibrium
ii. $\rho_s > r$, education market is an under-investment
iii. $\rho_s < r$, education market is an over-investment

The relationship between ρ_s and r can be presented in the following figure:

Figure 6.1: The relationship between ρ_s and r

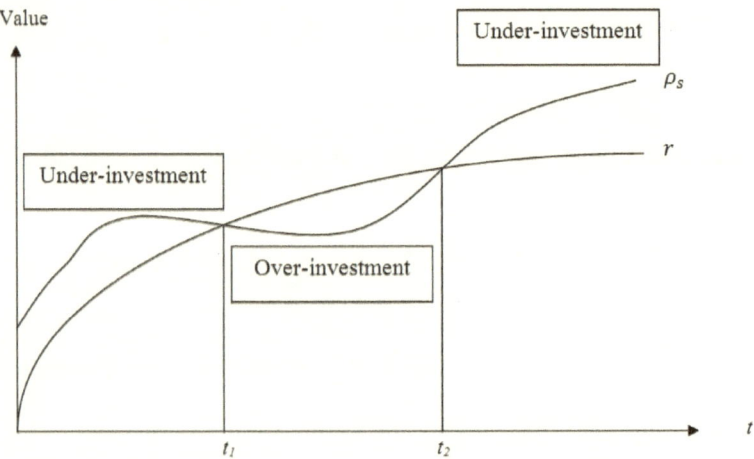

6.3.3. The model

It is assumed that agricultural households' productivity has the following form:

$$Y_{i,t} = A_{i,t} (Land_{i,t})^\alpha e^{\rho_s(Edu_i)+\alpha_{1i}x_{i,t}+\beta_{1i}Z_i+\mu_{i,t}+u_i} \quad (6.9)$$

where

- $Y_{i,t}$ is income of household i at time t.
- $Land_{i,t}$ is the amount of land of household i at time t, measured by square metres
- Edu_i is the education attainment of the head of household i. We might also employ this indicator for presenting average education attainment for a specific household in measuring the effect of education to the household's income, because the head of the household is always the main source of income in the family.
- $x_{i,t}$ is time-variant variables of the head of household i at time t which contains a log of expenditures on education, experience and experience square.
- Z_i is time-invariant variables of the head of household i which includes household size, latent variables that indicate whether the head of the household holds a primary, lower secondary, or upper secondary level of education.
- $\mu_{i,t}$ is individual unobservable effect
- u_i is the disturbance term
- ρ_s is an average rate of return of investment on the education of the head of the household
- α, α_{1i}, and β_{1i} are the effect of land, time-variant and time-invariant variables, respectively.

Taking log of both sides of equation (6.9) and rearranging it, we have:

$$\ln(Y_{i,t}) = \ln(A_{i,t}) + [\alpha \ln(Land_{i,t}) + \alpha_{1i} x_{i,t}] + [\rho_s Edu_i + \beta_{1i} Z_i] + \mu_{i,t} + u_i \quad (6.10)$$

If we denote $y_{i,t} = \ln(Y_{i,t})$, $a_{i,t} = \ln(A_{i,t})$, and ($Land_{i,t}$), equation (6.10) can be rewritten as follows:

$$y_{i,t} = a_{i,t} + [\alpha land_{i,t} + \alpha_{1i} x_{i,t}] + [\rho_s Edu_{i,t} + \beta 1_i Z_i] + [\mu_{i,t} + u_i] \quad (6.11)$$

Where the first bracket presents time-variant variables, the second bracket contains time-invariant variables and the last bracket shows the disturbance terms.

6.4. Methodology

Measuring the returns to schooling has to date attracted significant attention. There exists a plethora of discussions and arguments on estimating the potential relationship between individual ability and schooling. This is due to the correlation sometimes being negative when we treat schooling as an endogenous variable; however, a positive relationship may lead to a bias estimation when we use an OLS approach (Hausman & Taylor, 1981). According to Griliches (1977) and Griliches, Hall, and Hausman (1978), when we treat schooling as an endogenous variable and choose family background variables as instruments, the schooling coefficient usually increases by up to 50%.

The effect of investment on education in Vietnam's agricultural households is estimated by using Hausman and Taylor's (1981) model. The Hausman and Taylor estimator (hereafter HT) is a well-known method, as both authors provide a way for estimating that includes time-variant and time-invariant variables. By applying the HT model we can avoid an all-or-nothing approach between fixed and random effects in estimation. Applying the HT model in this chapter will have three benefits: (1) 'the ability to control for individual-specific effects—possibly unobservable—which may be correlated with other included variables in the specification of an economic relationship' (Hausman and Taylor 1981, p. 1377); (2) allows for estimating coefficients matching between time-invariant, but classifies the characteristics of farmers such as education, gender, region, sex and their earnings; (3) gives consistent results where some variables may correlate with various particular effects.

The HT model of unobservable individual effects, which has time-variant variables, x, and time-invariant variables, Z is written as

$$y_{i,t} = x'_{1it}\beta_1 + x'_{2it}\beta_2 + Z'_{1i}\gamma_1 + Z'_{2i}\gamma_2 + \alpha_i + \varepsilon_{i,t} \quad (6.12)$$

Where there is no correlation between the regressors with subscript 1 and α_i, there is a correlation among the regressors with subscript 2 and α_i. All regressors are assumed not to have any correlation with $\varepsilon_{i,t}$. The HT method can be seen as a type of random effect transformation that leads to the following model:

$$\tilde{y}_{(i,t)} = \tilde{x}'_{1i,t}\beta_1 + x'_{2i,t}\beta_2 + \tilde{Z}'_{1i}\gamma_1 + \tilde{Z}'_{2i}\gamma_2 + \tilde{\alpha}_i + \tilde{\varepsilon}_{i,t} \quad (6.13)$$

with, $\tilde{x}'_{1i,t} = \tilde{x}'_{1i,t} - \hat{\theta}_i \bar{x}'_{1i}$, and $\hat{\theta}_i$ being employed from the result of the regressing equation (6.12).

To measure unobservable individual effects of investment on education to agricultural households' income, as presented by head of household, we estimate the following equation:

$$y_{i,t}=a_{i,t}+[\alpha_{11} x_{1i,t}]+[\alpha land_{i,t}+\alpha_{12} x_{2i,t}]+[\beta_{11} Z_{1i}]+[\rho_s Edu_i]+[u_i+\mu_{i,t}] \quad (6.14)$$

where $x_{1i,t}$ and Z_{1i} are assumed not to correlate with u_i, while $land_{i,t}$, $x_{2i,t}$, and Edu_i are assumed having correlation with u_i.

6.5. Data summary

The datasets used in this chapter are the household-level data from three nationally representative surveys by the Vietnamese General Statistics Office (GSO) for 2002, 2004, and 2006. The Vietnam Household Living Standards Surveys of 2002, 2004, and 2006 (hereafter VHLSS 2002, 2004, and 2006, respectively), with sample sizes of 30,000, 9,000, and 9,000 (in that order) were useful datasets when researching the welfare or economic development at household level. All surveys were well-designed and matched international standards, as they were technically supported by international agencies such as the World Bank and the United Nations Development Programme.

6.5.1. Sample and variable definitions

The data was taken from the VHLSS 2002, 2004, and 2006. The sample contained households attending the survey in three waves. In each wave of the survey, some households were re-interviewed; thus, the VHLSS 2002, 2004, and 2006 are not true panel data. This analysis is based on agricultural households. These households were followed from

2002 to 2006. Only 937 households had working members in this panel data at the time of interview, mainly in agriculture. As such, they do not represent the entire sample of the VHLSS. The empirical results should therefore be carefully expanded to the entire country. Additionally, the time of interview should be treated more thoughtfully.

The VHLSS collects schooling and employment data as historical events. Schooling data includes the highest education certificate achieved by household members. These members are also asked whether they are studying at the time of interview. Employment data includes the main occupation of interviewees. Based on this data, the author selected individuals working primarily in agriculture. In three waves of survey, employment data was collected back to one year previously.

Years of schooling: to simplify the estimation of this variable, I considered two elements: (i) the highest certificates achieved by respondents, and (ii) respondents studying or who have studied at the time of interviewing. For the first element, I used the answer to the question *'What was the highest diploma [Name] obtained?'* and for the second element, I used answers to two questions: (i) *'What type of school is [Name] attended?'* and (ii) *'Is [Name] currently attending school?'* to estimate household members' years of schooling. The estimation was based on the length of study presented in Figure 6.1. There are three latent variables that indicate household members' educational attainments.

Work and employment: The work and employment assignment used data on work status over a year as reported by respondents. Respondents were considered to be working mostly in agriculture if they spent most of their working time in agriculture. These respondents were not attending school. Therefore, respondents working and studying at the time of the survey were dropped.

Income: total income of agricultural households earned from several sources of income. The total income is the net income of households, and equals output value minus expenditure. To estimate their total income, I used answers to the following questions: (i) *'What is amount in cash and in-kind obtained from land rental in the last 12 months?'*; (ii) *'What is the total amount you obtained from sales/barter for the last 12 months?'*; (iii) *'Value of the output harvested for the past 12 months'*; (iv) *'How much has your household spent on the following crops?'* Because some members of agricultural households have worked on some types of production, such as agriculture, livestock breeding, forestry, aquaculture, and providing agricultural extension services, the same process was applied for all types of production. If a household worked on only one or some listed business activities, total income from other types of business activities equalled nil.

Household size and experience: household size is the number of household members and experience is equal to household members' age minus years of schooling.

6.5.2. Descriptive statistics

Table 6.2 shows a summary of all variables employed to estimate rates of return to investment on education. There are, as noted, 2.811 person-periods in the dataset.

Table 6.2. Statistical summary

Variable	Number of observation	Mean	Standard Error	Min	Max
Natural logarithm of income	2811	9.22	1.00	3.17	12.94
Natural logarithm of land	2808	8.86	1.39	3.40	16.54
Natural logarithm of education expenditure	2144	6.42	1.13	2.30	9.96
Experience	2811	35.62	13.86	9	85
Experience square	2811	1461.20	1165.07	81	7225
Schooling	2811	6.77	3.28	0	12
Household size	2811	3.27	1.32	1	9
Primary (=1 if graduate primary education only; =0 other wise)	2811	0.37	0.48	0	1
Secondary (=1 if graduate lower secondary education only; =0 otherwise)	2811	0.50	0.50	0	1
High (=1 if graduate upper secondary education only; =0 otherwise)	2811	0.12	0.33	0	1

Source: Author's estimation from the VHLSS 2002, 2004, and 2006

As can be seen from the above table, the average years of schooling of active farmers were 6.77 years. This means that the major part of farmers left school one or two years after graduating primary education. This implies the low level of educational attainment of agricultural workers, which in turn implies the low income level of farmers. Theoretically, the more education people receive, the higher income they earn. The notion that arises in this context is that the modern economy requires more knowledge and skills for adapting and as such, farmers must apply new techniques to their work if they find themselves lacking capacity to do their jobs.

The average years of schooling varied between regions. The Red River Delta was reported as the region with the highest average, while the Mekong

River Delta was the lowest. Farmers in the Red River Delta attended an average of 8.15 years in schools, whereas farmers in the Mekong River Delta spent an average of only 4.78 years in school. The difference between the highest and lowest region was 3.37 years. Interestingly, the average income of farmers in the Red River Delta was lower than the mean income of farmers in the Mekong River Delta. Therefore, the effect of education attainment of farmers on their income in different regions did not reflect the theoretical relationship between income and education very well. Detailed statistical information by region is presented in the following table.

Table 6.3. Statistical description of income and education among regions

Variable	Red River Delta	North East	North West	North Central Coast	South Central Coast	Central Highlands	South East	Mekong River Delta
Income*	9.04	9.11	9.19	8.97	8.78	9.62	9.52	9.98
	(0.83)	(0.83)	(0.90)	(0.88)	(0.93)	(0.92)	(1.37)	(1.14)
Schooling	8.15	7.02	5.14	8.13	5.79	5.33	6.18	4.79
	(2.66)	(3.04)	(3.69)	(2.99)	(3.24)	(3.47)	(2.71)	(2.91)

Source: Author's estimation from the VHLSS 2002, 2004, and 2006
Note:
- *: Natural logarithm of income.
- Numbers in parentheses are standard error.

As Table 6.3 shows, farmers in the dataset had an average of just over 35 years of experience working in agriculture. This indicates that they have clear knowledge about the characteristics of their crop land and the type of products and inputs for their harvest. This indicator implies that farmers worked largely based on experience. This way of working is claimed mainly by provincial authorities to the central government in instances where they could not apply new technology to agriculture on a large scale. Additionally, looking at data on household size, the average number per household was slightly more than three persons. This indicates that, for the average dataset, each family consisted of parents and one child. For a family this size and considering the current level of education of Vietnamese farmers, it is very difficult to transfer new technology to agriculture. In fact, it is a limitation of agricultural development in Vietnam. Therefore, considering this point, we can begin to formulate a policy implication for the government.

6.6. Empirical results

First, we will estimate equation (6.6) for years 2002, 2004, and 2006 in order to gain an initial view of the effect of schooling on the incomes of agricultural households, as represented by the heads of households. Both estimations were applied using traditional OLS and GLS estimators, respectively. Secondly, OLS and GLS were employed to estimate equation (6.6) but with a smaller sample, where the focus was on households who participated in the two surveys. Thirdly, the Hausman and Taylor method was applied to estimate equation (6.14) and is reported in Table 6.7.

As can be seen from Table 6.4, the effect of land on income varied dramatically due to the differences between the two samples; however, the value of coefficients expresses the real relationship between land and agricultural households' income. In 2002, land had a strong effect on income, while in 2004, the effect of land on income was lower. Land coefficient in 2006 was larger than in 2004 but still lower than in 2002.

Experience and experience squared were not particularly statistically significant, but the signs of both variables were appropriate to Mincerian's earnings equation. The insignificance of experience and experience squared coefficients suggests that there is an inconsistency with the contemporary economic theory of earnings. Both Becker (1993) and Mincer (1974) agree that experience has a significant effect on earnings. Although the research of these two prominent authors focuses on the industrial sector, the same conclusion should be applied for agricultural households. We will come back this issue when the results of applying the Hausman and Taylor method are presented.

Educational expenses are an important variable for showing the link between two generations. In the agricultural sector, farmers in Vietnam do not pay more money for further training because their income is inadequate for this type of investment. However, agricultural households *are* paying for their children's education. Attending school at the primary level is compulsory, but doing so at the secondary level is still not compulsory. Farmers therefore have two choices: (i) send their children to school, or (ii) do not send their children to school. For statistics on enrolment, the two surveys cannot serve us properly; therefore, we need to use other statistics sources. These are provided in appendix A, Table A.2. The positive sign of the educational expenses coefficient reflects the affirmative link between two generations. Moreover, the positive sign of the coefficient represents the desire of parents for their children to have enough resources to attend school. The larger the value of the coefficient is, the greater the chance of children attending school is. It is also perceived that agricultural households are poor households; as such, the current generation will want to prevent the next generation form being as poor as they are. The only way to escape poverty is through investment in human capital, especially in education.

Table 6.4. Determinants of agricultural households' income (cross-sectional sample)

Dependent variable: Log income

Year Variables	2002 OLS	2002 GLS	2004 OLS	2004 GLS	2006 OLS	2006 GLS
Land	0.718***	0.757***	0.207***	0.156***	0.308***	0.359***
	(0.007)	(0.007)	(0.013)	(0.013)	(0.010)	(0.010)
Experience	0.0018	0.0025	0.009	0.0093*	−0.0119**	−0.0113**
	(0.004)	(0.003)	(0.0056)	(0.0054)	(0.0058)	(0.0053)
Experience square	−0.00003	−0.00005	−0.00005	−0.00005	0.00015**	0.00012*
	(.00007)	(0.00006)	(0.00009)	(0.00008)	(0.00007)	(0.00006)
Households size	0.011**	0.011**	0.087***	0.097***	0.0345***	0.0364***
	(0.005)	(0.005)	(0.007)	(0.0064)	(0.0109)	(0.0103)
Log educational expenses	0.072***	0.075***	0.144***	0.125***	0.126***	0.0944***
	(0.008)	(0.0075)	(0.0114)	(0.0104)	(0.0143)	(0.0140)
Schooling	0.022***	0.010***	0.055***	0.0696***	0.0109**	0.0218***
	(0.0033)	(.0031)	(0.005)	(0.0054)	(0.0046)	(0.0045)
Constant	2.267***	1.958***	6.313***	6.734***	5.769***	5.465***
	(0.089)	(0.093)	(0.147)	(0.147)	(0.163)	(0.1607)
Adj - R^2	0.6253				0.1955	
R^2		0.6244	0.3075	0.2985		0.1932
rho		0.193		0.145		0.1882
No. of observation	7172	7172	2503	2503	4277	4277

Note: - *, **, and *** denote statistical significance at 10%, 5%, and 1%, respectively
- Standard Errors are in parentheses.

Source: Author's estimation

Our main variable of concern in this instance is schooling. The value of the schooling coefficient is suitable with microeconomic theory and labour economics. The positive sign and value of schooling in both OLS and GLS estimators means that schooling has significantly contributed to agricultural households' income. However, when we compare values of the coefficient with interest rate for the same year, we find that $\rho_s < r$[36]. Thus, Vietnamese farmers are a case of over-investment in education of agricultural households. This is obviously wrong in the case of Vietnamese agricultural development, because education in Vietnam at that time should instead be a case of under-investment. Clearly, a case of over-investment can only happen in developed nations such as OECD countries, where education institutions are abundant and people have easy access to the services provided. Moreover, in the case of Vietnam, over-investment in education may only happen to the affluent class and absolutely not in the case of poor households such as farmers. Consequently, the estimation results here are true, but may not reflect the reality of economic theory.

---

[36] According to ADB (2010), the external interest rates in 2002, 2004, and 2006 were 7.8, 7.56, and 8.4 percent per year, respectively.

Table 6.5. Determinants of agricultural households' income (panel sample)

Dependent variable: Log income

Variables	2002		2004		Panel 2002 and 2004	
	OLS	GLS	OLS	GLS	OLS	GLS
Land	0.641***	0.683***	0.607***	0.606***	0.592***	0.618***
	(0.015)	(0.017)	(0.0201)	(0.021)	(0.022)	(0.014)
Experience	−0.0014	−0.0005	0.0037	0.0034	0.015**	0.0104*
	(0.007)	(0.0062)	(0.0086)	(0.0085)	(0.006)	(0.0054)
Experience square	0.00002	0.00002	−0.00011	−0.0001	−0.0002**	−0.00016*
	(0.0001)	(0.0001)	(0.0001)	(0.0001)	(0.00009)	(0.000083)
Households size	−0.0033	−0.005	0.035***	0.035***	0.0136	0.018**
	(0.0104)	(0.010)	(0.013)	(0.013)	(0.0095)	(0.0087)
Log educational expenses	0.074***	0.074***	0.098***	0.108***	0.139***	0.119***
	(0.016)	(0.016)	(0.0187)	(0.019)	(0.014)	(0.013)
Schooling	0.063***	0.035***	0.0295*	0.028*	0.0291***	0.025**
	(0.013)	(0.012)	(0.0153)	(0.015)	(0.0108)	(0.0102)
Primary	0.239**	0.142*	0.452***	0.441***	0.325***	0.285***
	(0.092)	(0.085)	(0.119)	(0.117)	(0.078)	(0.076)
Lower Secondary	0.153**	0.107*	0.319***	0.310***	0.226***	0.203***
	(0.065)	(0.060)	(0.086)	(0.085)	(0.053)	(0.054)
Upper Secondary	0.033	0.016	0.031	0.038	0.019	0.032
	(0.053)	(0.049)	(0.0703)	(0.069)	(0.042)	(0.044)
Constant	2.412***	2.235***	2.903***	2.851***	2.503***	2.486***
	(0.215)	(0.212)	(0.275)	(0.279)	(0.231)	(0.185)
Adj - R^2	0.5531		0.4213			
R^2		0.5508		0.4246	0.4253	0.4239
rho		0.191		0.041		0.188
No. of observation	1450	1450	1470	1470	2920	2920

Note: - *, **, and *** denote statistical significantly at 10%, 5%, and 1% respectively
- Standard Errors are in parentheses.

Source: Author's estimation

Secondly, equation (6.1) was estimated using panel data from 2002 and 2004. This panel data has a smaller number of observations, but can nonetheless help us to determine a closer result for repeated households in the three surveys. As can be seen from Table 6.5, the values of land coefficients contributed significantly to households' income in 2002 and 2004, respectively. For small farmers, land is still important to their income, while those with larger areas of land depend on it less. The land coefficient for 2004 in the panel sample was much larger than in the cross-sectional sample, as presented in Table 6.4. Experience and experience squared were statistically significantly in this instance and the signs of both variables were suitable with economic theory of education when we estimated equation (6.6) by using the panel sample. We can see in Tables 6.4 and 6.5 that experience and experience squared were not statistically significant when we estimated the equation in a cross-section sample. In contrast, the sizes of household coefficients in the panel sample estimation had lower value than the cross-section sample. Estimations of household sizes in the panel data were not as significant as in the cross-section data. This may lead to the conclusion that family background does have a significant effect on the income of households over a short period of time.

Years of schooling contributed significantly to agricultural households' income, not only in the specific year, but also over a longer period of time. With a smaller sample and in panel data, where households were re-interviewed over two years, years of schooling had a strong effect on the earnings of farmers. Again, however, the values of schooling were lower than the interest rate, so it is consistent with above conclusion. The effects of schooling on income were clearer when we employed three dummy variables. The dummy variable *primary* has value 1 for head of households who graduated primary school only, and 0 for the other; the dummy variable *lower secondary* has value 1 for head of households who graduated lower secondary school and 0 for the other; the dummy variable *upper secondary* has value 1 for head of households who graduated upper secondary and 0 for the other. Only primary and lower secondary variables were statistically significant, while upper secondary variables were not significantly statistical. This means that for farmers, primary and lower secondary schooling is necessary in the context of Vietnamese agriculture. If we did not place importance on this statistical significance, we can see that the effect of primary schooling on income was higher than lower secondary schooling, with the effect of upper secondary schooling being the lowest.

Additionally, equation (6.14) was estimated by using the Hausman and Taylor method and presented in Table 6.6. Hausman and Taylor suggest that their method provides a better way for estimating the rate of return of schooling, and that the coefficient will be higher than the traditional method. It is expected that the outcomes of estimation using the VHLSS dataset will follow Hausman and Taylor's direction accurately; as such, we can conclude that the nature of investment in education in Vietnam is under-investment. This conclusion is more appropriate for the case of Vietnam.

In the context of applying the Hausman and Taylor method, the effect of years of schooling on agricultural income increased significantly. The minimum value was 0.219 and the maximum value was 0.324. Clearly, the minimum value of this coefficient is larger than the interest rate, so this result should be seen as a more reliable estimation of the rate of return of investment in education. Based on empirical results, one more year of schooling will increase farmers' income at least 21 percent. Over a longer period of time, from 2002 to 2006, the effect of schooling on farmers' earnings was higher than over a shorter time, from 2002 to 2004 – 0.2591 compared to 0.219.

Table 6.6. Contribution of investment in education to agricultural households' income

Dependent variable: Log income

Variables	HT/IV (1)	HT/IV (2)	HT/IV (3)	HT/IV (4)
Time variant exogenous variables				
Land	0.181*** (0.021)			
Experience	0.245*** (0.0167)	0.262*** (0.0168)	0.0583*** (0.0119)	
Experience square	−0.0025*** (0.00027)	−0.0027*** (0.0003)	−0.00042*** (0.00014)	
Log educational expenses	0.084*** (0.0164)	0.083*** (0.016)	0.1833*** (0.0223)	0.1972*** (0.0235)
Time variant endogenous variables				
Land		0.126*** (0.021)	0.208*** (0.0127)	0.1654*** (0.0139)
Experience				0.1222*** (0.0175)
Experience square				−0.00093*** (0.00021)
Time invariant exogenous variables				
Households size	0.090** (0.043)	0.132*** (0.0461)	−0.0564* (0.0335)	−0.1071*** (0.0382)
Primary	1.485* (0.859)	1.126*** (0.308)	4.095*** (0.4544)	2.6631*** (0.2181)
Lower Secondary	0.511 (0.357)	0.547** (0.268)	2.887*** (1.072)	1.2663*** (0.1273)
Upper Secondary	0.536 (0.369)	0.330 (0.232)	2.102 (1.469)	

Variables	HT/IV (1)	HT/IV (2)	HT/IV (3)	HT/IV (4)
Time invariant endogenous variable				
Schooling	0.324** (0.158)	0.219*** (0.0392)	0.2591** (0.1323)	0.3382*** (0.0271)
Constant	Yes	No	No	No
Rho	0.9745	0.9783	0.7824	0.8048
No. of observations	2920	2920	2142	2142
Years	2002, 2004	2002, 2004	2002, 2004, 2006	2002, 2004, 2006

Note: - *, **, and *** denote statistical significantly at 10%, 5%, and 1% respectively
- Standard Errors are in parentheses.

Source: Author's estimation

In the first column of Table 6.6, land is treated as a time variant exogenous variable and schooling has the largest effect. Among the three dummy variables indicating the level of graduation among heads of agricultural households, only *primary* was statistically significant at 10 percent, while the two others were statistically insignificant. When land was treated as a time variant endogenous variable, both primary and *lower secondary* were statistically significant at 5 percent and 1 percent, respectively. However, the upper secondary dummy variable remained statistically insignificant. This result and the empirical results presented above suggest that for the case of Vietnamese agriculture, *upper secondary* graduation does not have a significant effect on the income of agricultural households. Particularly, in the panel of years 2002, 2004, and 2006, the effect of the primary variable was the largest.

Experience and experience squared were statistically significant at 1 percent and the effect of these variables on agricultural households' incomes was considerable. In the case of agriculture, experience should have a large effect on agricultural household incomes, because harvests depend greatly on vintage knowledge of the characteristics of the land, such as climate, rainfall, and the way of using fertilisers. Both experience and experience squared have appropriate sign, as suggested in Mincer (1958, 1974) and Becker (1966).

Educational expenses have a considerable effect on the earnings of farmers. This means that farmers are willing to pay for the education of their children and expect that the next generation will have a better economic future and useful skills. Having good skills is a positive for their children when they eventually join the labour market. The larger value of this coefficient illustrates the strong relationship between the current and next generation.

In the last column of Table 6.6, where we let land, experience, experience squared and schooling correlate with u_i, the result is robust. The coefficient of schooling has increased significantly, from 0.25 to 0.33, and the standard of error has decreased from 0.13 to 0.03. The assessment of agricultural households concerning spending on education for their children, as presented by educational expenses, is higher. The value of the coefficient of log educational expenses rose moderately from 0.18 to 0.19, but the standard error also increased slightly from 0.02 to 0.03. The signs of experience and experience squared variables are suitable with economic theory and their values are robust. This means that farmers consider experience as an important factor of their income.

In order to see the variation between geographical regions in rates of return to educational investment, equation (6.14) was estimated by using the HT/IV procedure for eight regions, which are reported in the following table.

Table 6.7. Rates of return on educational investment across regions in Vietnam

Variables	Red River Delta	North East	North West	North Central Coast
Time variant exogenous variables				
Log educational expenses	0.0247 (0.0406)	0.1569*** (0.0447)	0.1919 (0.1176)	-0.0485 (0.0563)
Time variant endogenous variables				
Land	0.3311*** (0.0280)	0.0751*** (0.0234)	−0.0222 (0.0634)	0.0897*** (0.0332)
Experience	0.1445*** (0.0386)	0.1311*** (0.0286)	−0.00319 (0.0746)	0.1464*** (0.0399)
Experience square	−0.00078* (0.00042)	−0.00036 (0.00039)	0.00176* (0.0010)	0.000085 (0.00049)
Time in-variant exogenous variables				
Households size	−0.2278** (0.1014)	−0.2728** (0.1063)	−0.3335 (0.2298)	−0.6087*** (0.1970)
Primary	0.9266* (0.5428)	2.8071*** (0.4182)	5.0523*** (0.9331)	1.9007** (0.7631)
Lower Secondary	0.7737*** (0.2687)	1.5801*** (0.3636)	2.5419*** (0.8383)	1.5275*** (0.5102)
Time in-variant endogenous variables				
Schooling	0.2949*** (0.0579)	0.4111*** (0.0611)	0.6626*** (0.1292)	0.5200*** (0.0844)
	South Central Coast	Central Highlands	South East	Mekong River Delta
Time variant exogenous variables				
Log educational expenses	0.3733*** (0.0709)	0.2011* (0.1063)	0.3377*** (0.1243)	0.2260*** (0.0633)
Time variant endogenous variables				
Land	0.2389*** (0.0508)	−0.00226 (0.0656)	0.0840 (0.0828)	0.3448*** (0.0597)
Experience	0.1036* (0.0577)	0.2205** (0.1063)	0.1637* (0.0966)	0.1102** (0.04475)
Experience square	−0.0010 (0.00065)	−0.0025* (0.00139)	−0.00138 (0.00129)	−0.000896* (0.000499)

Variables	Red River Delta	North East	North West	North Central Coast
Time in-variant exogenous variables				
Households size	−0.0959	−0.0232	−0.1823	−0.0606
	(0.1427)	(0.1082)	(1.6445)	(0.2390)
Primary	1.7498**	2.8680**	1.6527	1.8459*
	(0.8318)	(1.2591)	(4.3995)	(0.9513)
Lower Secondary	1.1021**	1.2774*	0.9895	1.2815
	(0.5332)	(0.6844)	(4.3381)	(1.0797)
Time in-variant endogenous variables				
Schooling	0.1823*	0.4030***	0.3557	0.2534
	(0.0970)	(0.1262)	(0.8355)	(0.1811)

Source: Author's estimation using VHLSS 2002, 2004, and 2006

The rates of return of investment on education varied among geographical regions and not all the coefficients were statistically significant. Farmers on the South Central Coast had the lowest rate, 0.18, while farmers in the North West had the highest rate, 0.66. Farmers on the North Central Coast had the second highest rate, 0.52. This result implies that farmers across regions value their educational attainment differently. The differences between regions may be as a result of the different perceptions of people in these regions. Farmers on the North Central Coast might consider education attainment more important than farmers in the Mekong River Delta as a result of their geographic-specific culture. Moreover, in regions lacking favourable conditions for agriculture, farmers often had higher rates of return of investment in education, except on the South Central Coast. Farmers in the North West had the highest rate of return, despite the natural conditions of the region not being good for agricultural development. The difference in coefficient of schooling implies that one year of further education will result in a difference in farmers' income across regions in Vietnam. Additionally, the coefficients of schooling in the South East and Mekong River Delta were not statistically significant, while the coefficients of other regions were statistically significant. Although these coefficients were not statistically significant, their sign and values still support economic theory on educational investment.

Regarding education level, farmers in poor regions showed a robust difference in education levels. The different levels were the highest in the poorest regions, that is, the North West. In these areas, a farmer with a primary or lower secondary education had a higher income than a farmer who did not graduate from school.

The values of experience coefficients varied slightly across regions. Excluding the negative value in the North West, the lowest value was 0.10 and the highest was 0.22. The positive values of these coefficients imply an increased relationship between the experience and income of farmers. Logically, it could be said that the older farmers are the higher income they earn; practically, however, older farmers are less productive than younger farmers. Older farmers have better experience, but younger farmers have better health. Therefore, when we look at the sign and value of the experience squared coefficient, its negative value implies that as farmers become older their productivity should decrease over time, leading to their incomes being lower than that of young farmers. This statement may not be applicable at a global level, but it may right for a specific case, i.e., Vietnam.

6.7. Policy implications

Empirical results suggest that farmers' past years of schooling had a significant effect on their productivity, as presented by their earnings. However, these farmers had been trained a long time ago and need to be retrained. Currently, farmers gain their knowledge from school and in the form of vintage knowledge. Vintage knowledge of farmers has two components: knowledge from school and knowledge from learning-by-doing. Therefore, the government should have an appropriate policy for encouraging the vintage knowledge of farmers in order to facilitate an increase in the income of farmers. Moreover, the Vietnamese government should construct a collection of experienced farmers from whom other agricultural households can learn. All types of communication should be used to contain and transfer knowledge, such as newspapers, newsletters, and documentaries.

More primary and lower secondary schools should be put in place for people in rural areas to avoid the illiteracy of minor ethnic groups or people in remote areas. For the current stage of agricultural development in Vietnam, farmers need to be trained at primary and lower secondary schools, which there are not enough of. Additionally, farmers are not only expected to have basic knowledge, but also to learn new scientific knowledge with the intention of improving their harvests and to apply new technologies to their crops during the current period of agricultural industrialisation. The Vietnamese government should therefore not only focus on general schooling, but also on providing farmers with opportunities for learning throughout their lives. Additionally, vocational training should be addressed, as farmers need to be trained in new skills so they can work for factories or companies in the areas where they live.

Education is very important to economic growth. As such, in addition to private investment in education, the government should also invest more in education. Government investment should focus on building more schools and enhancing the quality of teachers, such as capacity building for trainers at the schools. The authorities might increase the quality of schooling by applying new methods of training that focus on trainees, rather than making trainers central to the education process. This means that the Vietnamese government needs to ensure that the structure of educational levels is consistent and efficient. In this way, the education system will be able to provide learners with good knowledge.

6.8. Conclusion and further research

In this chapter, I applied Hausman and Taylor's method for estimating the rate of returns of investment in education. Based on empirical results, it is said that the outcome of the econometrics model supports the method that Hausman and Taylor introduced in their famous paper in 1981. The value of schooling coefficient is suitable with contemporary microeconomics and labour economics, especially the Mincerian earnings equation. Moreover, the estimated value of schooling expresses a realistic picture of Vietnamese agricultural development when the value points out those Vietnamese farmers are an under-investment case, unlike traditional methods such as OLS and GLS, which provide an over-investment case. From empirical results, several public policies have been built up. It is suggested that, for the case of Vietnamese agricultural development, Hausman and Taylor's method is better than traditional econometrics methods.

In this research, the effect of the industrialisation process was not caught in the model due to a lack of data. The effect of the industrialisation process on agricultural development, especially on farmers living, is currently a hotly debated topic in Vietnam. Therefore, further research is suggested on the effect of industrialisation on the earnings of farmers in a bid to establish appropriate public policy.

Chapter 7

SOCIAL CAPITAL AND AGRICULTURAL DEVELOPMENT IN VIETNAM

7.1. Introduction

Agriculture has developed alongside Vietnam's reform process of nearly 25 years. The reform process was initiated with Resolution 10 and the improvements it brought to the agricultural sector. There is a consensus among Vietnamese and international researchers that agricultural development of Vietnam is sourced by institutional change (Che, Kompas, & Vousden, 1999, 2006; Kompas, 2004; Kompas, Che, Nguyen, & Nguyen, 2009; Ravallion & Walle, 2008; T. Tran, Grafton, & Kompas, 2009). These groups usually explain that prior to the reform, the productivity of agriculture in Vietnam was low and Vietnam needed to import rice; however, following the reform, Vietnam became one of the largest rice exporters. The question is why did this event happen? The amount of land, the technical level of farmers' skills and natural conditions remained almost unchanged, while productivity differed greatly between, before, and after reform. The agricultural development of Vietnam has become an interesting example for the effect of institutions on economic development for courses on economics at many international universities. However, the question is ultimately answered at the macro level, while at the micro level there remains very little research focused on or responding to the small scale.

In Vietnam (similar to China), economic reform has positively affected agricultural development. 'De-collectivizing agriculture [and] implementing the household responsibility systems in farming' (Henin, 2002) has stimulated agricultural development and farmers' living standards in rural areas and the agricultural sector. Vietnamese political leaders who supported renovation agreed that collectivising agriculture has been unsuccessful and therefore further reform is needed such as liberalisation of agricultural production and privatisation of agricultural land. On the other hand, traditionalists point out that liberalisation of agriculture will lead to a high rate of unemployment in rural areas and as a result increase migration from rural to urban areas. Conservatives express concern that community power will deteriorate and call for protection from the Vietnamese government.

This chapter will focus on the investigation of the effect of social infrastructure on the quality of life of agricultural households and agricultural firms' productivity. Based on empirical results, policy implications will be suggested. The pro-reform policy should be implemented in order to achieve further success in the agricultural sector.

7.2. Background information on social capital related to agricultural development in Vietnam

7.2.1. Institutional change in Vietnam

In the mid-1980s, the Vietnamese government began an important strategy called *Doi moi* ('renovation'), which was initiated in the agricultural sector. The change in ideology occurred because of the food scarcity crisis and the need for bringing about change in communities. Consequently, agrarian reforms were implemented and led to a transformation in agriculture (Fforde, 1991; Benedict J. Tria Kerkvliet, 1993).

In April 1988, the Vietnamese Communist Party issued a vital policy, Resolution 10, called 'Renovation in Agricultural Management'. This was the first time agricultural households had been treated equally to collective or state sectors in rural areas. In all regions in Vietnam, farmers had been allocated, amongst other things, land and draught animals. Agricultural households decided what should be harvested on their land and were free to use their land for working towards better incomes. In fact, the state owned the land, including agricultural land and the communes and villages were in charge of contributing land to villagers under the administration of local authorities. Later, the transformation of agriculture was improved by the implementation of Land Law 1993, which pointed out that farmers have five specific rights where their land was concerned: exchange, transfer, rent, inherit, and mortgage land use rights. Additionally, the law indicated that farmers have limited time for using land for harvesting, such as 20 years for cropland and 50 years for aquaculture, and that their land rights could be renewed when their contracts expired.

The combination of the *Doi moi* policy and Land Law system has created a background for agricultural development. Land is being used effectively. By using contract and procurement procedures, land that was used only marginally before in collectives was put back into the production process. Consequently, a considerable amount of money was attracted from rural households, which was invested in agriculture. Since then, agricultural production has increased considerably. For example, rice production has risen from 18 million tonnes in 1987 to just over 25 million tonnes in 1996. From a country needing to import rice, Vietnam became an exporter in rice following only two years of transformation.

One more factor that can be seen as having had a policy change effect on agricultural production was the abandonment of the two prices system. As a result, farmers were encouraged to invest more in their land and crops. By eliminating the two prices system, the central planning mechanism in Vietnam ended. Consequently, the incomes of agricultural households increased significantly (Henin, 2002). Farmers can now buy necessary items for their cropping and can invest more money in their children's education. They are able to spend more money on their health and their savings have been increased.

De-collectivisation was a significant transformational aspect in Vietnamese agriculture. For a long time, collectives dominated Vietnamese agriculture and constrained its development in terms of economic productivity, while expanding on the side of social services (Beresford, 1990; Ravallion & Walle, 2008). The noteworthy amendment of the policy was the re-allocation of agricultural land. Members of collectives were allocated both good and low quality sections of land in order to apply equalisation. Alongside this policy, Vietnam has a free market for land-use rights, while the Vietnamese government retains land ownership. This is different from Chinese agricultural reform, where collectives retain the right to allocate and set quotas for agricultural land. This policy raises costs for farmers, because they have to spend more time harvesting in fragmented land areas. Additionally, crop land is wasted because of making the border among plots and the inability to use tractors.

The transformation of the Vietnamese government from a director to a rule-maker during the *Doi moi* era is becoming more important than ever before. The transformation has demanded changes in the structure, tasks, and functions of Vietnamese government bodies and in the legal system. Managing the transformation has demanded cooperation and coordination across ministries and agencies in order to minimise confusion and achieve policy objectives.

The Public Administration Reform Master Plan was launched in the 1990s, which addressed decentralisation, defining functions and tasks, and modernising public administration services. Some of the plan's outcomes have been achieved, such as simplification of administrative procedures; on the other hand, the conversion from policy to implementation has been slow. According to the estimation of the World Bank (2009), simplifying administrative procedures has positively affected the villagers' lives.[37] The World Bank's report indicates that villagers are more able to find guidance information than before, while the complaints of respondents related to land and housing procedures focuses on the confusing of procedures.

---

[37] Detailed information is in Worldbank (2009, p. viii).

The results of the VHLSS 2008 survey have been confirmed by other survey at firm levels, Provincial Competitive Index (PCI) project. Firms participating in the survey reported that they found it easier to access government documents than previously and they did not require a private relationship with officials to obtain said documents.

7.2.2. Social infrastructure change in Vietnam

'Social infrastructure' is a term raised by Hall and Jones (1999), who refer to social infrastructure as institutions and policies that affect the private and social returns of investment activities. Social and private returns may differ due to various types of investment and activities proposed to the individuals' present interests. Additionally, social infrastructure could be classified in three categories (D. Romer, 2006). *First,* elements of government fiscal policy are accounted for. The two instruments for this category are tax treatment for investment activities and marginal tax rates for labour income. The conditions that affect private decisions are the *second* category. In places where crime is properly checked, economic activities will increase more than in areas where crime is left unchecked. Furthermore, if contracts are not monitored closely, adhere to the courts' interpretation of the law, or the contract is unpredictable, the long-term investment activities will be less attractive. *Finally,* various factors affect the process of rent-seeking activities run by the government. Bribery in public procurement procedures is one form of rent-seeking activity. Corruption in infrastructure projects could be seen as other important form of rent-seeking. Another form of this activity is lobbied action.

Since the *Doi moi* policy was implemented in 1986, there has been much change Vietnam's social infrastructure in. The most significant transformation has been the quality and availability of information related to investment decisions. IMF (2000) complained that the Vietnamese economy set an example of a situation where there existed a lack of information for doing business. The Economist Intelligence Unit's report on Vietnam in 1997 pointed out that there was no business service, including information provision, in the Vietnamese economy. The lack of information service has been confirmed by several firm surveys in Vietnam.[38] Although considerable institutional reform has been implemented since 2000, the weaknesses of providing sufficient information regarding the market remains (T. Tran et al., 2009).

The second important change in the social infrastructure of Vietnam has been the creation of a market for trading agricultural products. This is a further step following the granting of the five rights of farmers on using land. By employing markets, the Vietnamese government has encouraged farmers to accumulate agricultural land to improve the productivity of agricultural households. Large farmers have appeared and now contribute a vital role for supplying food to the market and exporting produce. Additionally, this policy has helped 'war heroes and martyrs' by providing good job opportunities and arable land for agriculture. Individuals who have contributed significantly to the Vietnamese revolution or were injured during the Independent war, or who are not able-bodied have also been helped by the policy. For these individuals, the Vietnamese government has created a special policy, because of their involvement in the war and

---

[38] For more detailed information on this issue, please see T. Tran et al. (2009).

because they did not have the same opportunities as others to attend school. These individuals can change the purpose of using agricultural land without any approval from the government, but should nonetheless inform local authorities first. This is a kind of social protection policy supervised by the Ministry of Labour, War Invalids, Social Affairs, Ministry of Finance, and the local People's Committees.

The third significant change has been a rise in the decentralisation of public spending on social services. While using the cooperative agriculture mechanism, the Vietnamese government was in charge of supporting members of collectives in healthcare and education expenses. Following de-collectivisation, farmers have to pay healthcare and educational fees. These expenditures make up a significant part of their incomes, as farmers' earnings are quite low in comparison with salaries earned in other sectors. Vietnam has experienced a significant drop in government spending on public health and education. Meanwhile, under the decentralised process, lower level governments—district and provincial—have been made responsible for facilitating healthcare centres, sanitation services, water supply, and essential drugs for patients. Communes' authorities are responsible for improving the quality and production of crops through investment, operation, and maintenance activities.

The relationship between predictable law and investment activities in Vietnam is positive at the provincial level. Obviously, the changes in law have brought about uncertainty for the firms, households and government officials who have to follow them. Thus, the key challenge of legal reform is to ensure that the transformation of the legal system in Vietnam is consistent, implementable, and well understood. A stable and predictable system of laws will give firms and households more confidence in investing. In fact, the PCI survey reveals that provinces with a more predictable system of laws gain more investment from firms; however, firms now report that economic and financial laws in Vietnam are unpredictable (Worldbank, 2009, p. x).

7.2.3. The improvement of understanding the relative position of farmers in Vietnam

Before *Doi moi* was started in 1986, agricultural activities were implemented in the collectives using an egalitarian distribution principle, which minimised inequality among farmers in the countryside (Nam, 2001). However, the percentage of farmers who participated in the collectives differed from the North to the South on historical grounds.[39] The percentage of farmers who joined the collectives in the North and Central Coast of Vietnam was 98% and 89%, respectively, while the percentage of agricultural households participating in the collectives in the South of Vietnam was only 6% (Pingali & Xuan, 1992). Although the Vietnamese economy has been run like a market economy since 1986, the traditional notion of equality has remained in famers' minds. Them (1997) argues that a particular farmer may be concerned about what other farmers think about their activities and relations. For example, farmers in rural areas tend to think that in events held by their neighbours, such as funerals, weddings, harvests, and construction, they should participate as a matter of duty. This may be attributed to a long history of surviving together within the

---

[39] Before 1975, the South of Vietnam was more market oriented than the North of Vietnam (Nam, 2001).

environment, where the support of local officials and government had previously been lacking. Additionally, individuals' perceptions of their position in the community are affected by the common knowledge and tastes of the community in which they live.

Carlsson et al. (2007), based on experimental research conducted on the Dong Tam commune (Binh Phuoc Province, Southern Vietnam) in 2002, concluded that if a particular household has a member who joins a local People's Committee, the household tends to cares more about their position in the community than those not part of such a committee. However, a large part of interviewers did not worry much about their social standing. This may be explained by the transformation of people's perceptions about social standing during the period of changes in Vietnam's economy (Carlsson, Nam, Linde-Rahr, & Martinsson, 2007).

In other poverty alleviation programs[40] run by a joint effort between Vietnamese and Swedish governments in Ha Giang and Quang Tri provinces, villagers were encouraged to make themselves heard by improving their knowledge about management and their rights to engage in planning process at district level. In an interview, chairman of the Suoi giang Commune's People Committee in the Van Chan district, Ha Giang province, stated:

> **The handing over of power** to the village is good because villagers know what they need, and what investment should be made in their village. One could say that their decisions are more informed. For example, in one program

---

[40] This programme is called 'Chia-se', meaning 'sharing' or 'partnership'.

the district decided that hoes and shovels should be granted to villagers. But most villagers already had them. So when exercising village empowerment, it is better to let people take the initiative in deciding what is the most necessary for them (MPI & Sida, 2006, p. 13).

This means that if power is transferred to villagers, their social capital will be raised and they will express such capital much more apparently.

Another aspect of the issue is the understanding of people concerning their rights of knowing information related to their needs and life. In the VHLSS 2008, nearly 50% of participating households reported that they did not have enough information on the budget and plans of communes. Thus, they did not have enough information to build up their own plan and contribute their opinions to the work of local authorities. Despite the fact that people ask local cadres for information on commune budgets and plans, their information requirements are not met. This lack of information has brought about hesitation on the part of people to make decisions, creating missed opportunities for increasing their incomes. The reasons for the lack of information can be listed as follows:

- Literacy and language issues
- Education
- The distances from living areas to the source of information
- Lacking a basis for communication and information infrastructure

7.3. Theoretical framework

7.3.1. The basic model of social capital in the context of agricultural development

A society may consider a society to be a nation, region or a community. The society has finite members, which include individuals and two types of institutions: (i) private firms, and (ii) social institutions that supply services of social capital. At the micro level, 'social capital is observed as being created by the formation of relationships, which in turn are determined by conditions on the individual level (personal traits, personal resource collections, and investment in relationships[41])' (Van der Gaag, 2005, p. 15). Indeed, investment in relationships is a way of investing in social relationships, in which individuals actively invest their personal resources with the aim of increasing their living standards. At the macro level, social capital refers to social infrastructure such as roads, highways, ports and public transport systems—the process of having access to social common capital and the use of social capital in order to raise the individual's benefits.

Social capital taxes are levied when individuals use it. These taxes are set by the government and may therefore differ between regions, societies, and nations. One crucial role of government is to determine the taxes of using social capital services in such a manner that the pattern of distribution of economic resources and incomes are at an optimum and well-defined. According to the economic theory of social capital, we must build up the

---

[41] Investment in creating relationships does not include negative actions such as corruption and bribery.

concept of social capital in an operational form and analyse the effect of investment on social capital at the macro and micro levels, on the living standards of people and on the performance of agricultural production of firms and households.

As discussed above, we have considered only the agricultural sector; however, 'agriculture' concerns not only industrial and economic issues but also almost all aspects of human life: cultural, social, and natural. It provides us with food and materials such as silk, cotton and wood that sustains our existence. Additionally, agriculture sustains the natural environment, including underground water, forests, surface water, wetlands, soil, and the atmosphere. Agriculture has provided grounds for 'a harmonious and sustainable relationship between nature and humankind through the social institution of the rural community in many East Asia countries' (Uzawa, 2005, p. 219). Thus, agriculture can be seen as a 'light' version of the economy as a whole.

Individuals are generally symbolised by $indi = 1,\ldots,i$, while villages in agricultural sectors are presented by $villages = 1,\ldots,v$ and firms are denoted by $firm = 1,\ldots,f$. It is assumed that villagers and firms produce agricultural products by using two types of production factors: variable and fixed factors. Fixed factor households are presented by a square of land, whereas variable factor households are presented by different types of labour. In this chapter, only one kind of labour in the agricultural sector is assumed. For firms, variable and fixed factors are quite different and more complex than for agricultural households.[42] Agricultural products of households are denoted by $house\_product = 1,\ldots,z_1$ and agricultural products of firms in

---

[42] To understand more about variable and fixed factors of firms, please read textbooks on corporate finance.

the agricultural sector are presented as *firm_product* = $1,...,z_2$, and $z_1 \neq z_2$. The total goods consumed by a particular individual will be denoted by a composition of *house_product* and *firm_product*, therefore consumption^{indi} = (*house_product, firm_product*).

It is assumed that a particular individual will invest in specific types of relationships[43] through their decision of joining the labour market. Farmers who work for themselves on their cropland will have different investment activities than farmers who intend to work for bigger farms. When farmers decide to work for themselves, they directly serve the market and then build up their relationship with their customers. Thus, their social relationships will be classified according to two types: (i) relationships with input suppliers, i.e., supply-side; (ii) relationships with output customers, i.e., demand-side. On the other hand, farmers who decide to work for big farms or industrial organisations will consider building up their social relationships with their colleagues. Both types of farmers will have different investment strategies in social relationships in order to achieve their own targets, based on their available financial and non-financial resources (for further analysis, please see Dufhues, Buchenrieder, & Fischer, 2006).

The return of investment in social relationships is yielded in three possible varieties: (i) economic return is presented by financial or

---

[43] It is assumed that agriculture has three types of formal institutions (firm, government, and social organisation), so there are three sources of building up social capital: (i) firm source; (ii) government source; (iii) social organisation source. Each source has its own characteristics and requirements for joining. For example, firms require their members to have creative and active minds, while governments require their members to have stable and less creatively-inclined minds.

non-financial benefits or both; (ii) political return is presented by the good chance of promotion in a hierarchical system; (iii) social return is presented by the increasing of reputation of an individual in the network. Additionally, investment in social capital might result in better physical health, mental health and life satisfaction. For physical health, the return includes freedom from illnesses and harm, while for mental health, 'the return reflects [the] capability to withstand stress and maintenance of cognitive and emotional balance' (Van der Gaag, 2005, p. 16). Life satisfaction indicates optimism and contentment about life.

7.3.2. The effects of social infrastructure on households' income

Social infrastructure is defined as 'the institutions and government policies that determine the economic environment within which individuals accumulate skills, and firms accumulate capital and produce output' (Hall & Jone, 1999, p. 84). Social infrastructure can therefore be categorised into two components: institutions and government policies. Each element of social infrastructure has affected individuals and firms, depending on the characteristics of said individuals and firms. For example, institutions and government policies may affect the opportunities of an individual to acquire essential skills, while these policies might affect the chances of a firm in accessing credit sources or the labour market. Certainly, the interaction between individuals and firms with institutions and government policies works two ways. Institutions and government policies form the rules of the market, whereas individuals and firms cooperate with institutions and government bodies in issuing and implementing public policies.

Social infrastructure has contributed to the welfare of individuals by providing facilities for raising individuals' incomes. The rise of agricultural household incomes will lead to the raise of farmers' welfare, usually presented by their total expenditures. The welfare of an individual is presented by his/her utility function. If we define *U(indi)* as the utility function of individual *i*, we have

$$U^{indi} = U^{indi}(house\_product, firm\_product, leisure, relation, house\_characters).$$

This means the utility of a particular agricultural household is a function of goods that the household uses for living, and t is leisure time, and the creation and maintenance of social relationships, which reflects household tastes in purchasing goods and services. In terms of social infrastructure, local markets and local roads can be seen as the availability of social infrastructure. The availability of a local market with an adequate supply of good food and other necessary goods will satisfy the utility of individuals and households. The availability of a local market also increases the chance of agricultural households meeting with their customers and their opportunities for gathering the necessary goods for their production and consumption.

Households' utility has faced budget constraints, *c = p * house_product + w * labour_supply + r * socialized*, where *p* represents the prices of selling household products in the market, *w* is the wage of working in the labour market and *r* is the benefit gained from creating and maintaining relationships.[44] Clearly, individuals

---

[44] Here, the relationship excludes several types of relationships that are closed to corruption or bribery.

cannot directly measure the benefits of keeping in contact with their friends, relatives and other valued persons. The benefit of maintaining a relationship may instead appear when a particular individual has done something that he/she would not have been able to do without the help of friends, relatives, or others. This benefit may not be measurable in financial value, but can instead be estimated by the satisfaction or happiness that it yields, such as leisure time. However, keeping contact is not only for the purpose of maintaining a relationship or network but also a way of investing in the future. Joining a network or communicating with friends, relatives and other valued persons will yield more opportunities for gaining benefits. For example, the transformation of common knowledge among members of a network may enhance their opportunities for finding better employment, or discovering ways for solving an individual problem. Therefore, the time spent on maintaining social relationships should be seen as a current action that is expected to enable better incomes in the future.

If we define $E_{invest}(indi)$ as the expectation of social capital investment of individuals, we get: $E_{invest}(indi) = I(types\ of\ relationship)$, where $I(types\ of\ relationship)$ is investment in social relationship that may encourage households or individuals to earn more in the future. In truth, it is the availability of a social infrastructure that encourages the development of social relationships, which will increase the benefits and decrease the risks of investment in relationships. On the other hand, a lack of social infrastructure will constrain communication among members of a network, and as a result, the link between members will deteriorate. Therefore, $E_{invest}(indi)=I(types\ of\ relationship\,|\,availability\ of\ social\ infrastructre)$, and we have the following definition:

Definition: *Investment in the social capital of individuals or firms or other parties is an expectation of gaining future value through current spending on social relationships, the likes of which are constrained by the availability of a social infrastructure.*

In reality, when a group gains benefits for its members, the benefit of the other groups may be harmful. For example, car producers in Vietnam do not want to reduce the welfare of their customers, but are nonetheless willing to lobby the Vietnamese government to issue a policy that limits the quantity of imported cars, thereby gaining more benefits from customer expenses. The policy implication here is that we can predict if the welfare of a particular group's members will be raised when the group implements cooperative activities; however, the effects of the policy on the welfare of other groups are usually ambiguous.

Two communes in the same district may have different social capital, despite having the same rights for using the social infrastructure of the district. The differences in availability of Internet or telephone access may be a core reason for the inequality of villagers within these two communes. Indeed, in the same village, different families might have dissimilar social capital, because they have accessed the social infrastructure in different ways. For example, a rich family will be able to directly access the healthcare and education system because they can pay for using these services, while a poorer family needs support from local and central government for accessing basic needs services.

7.3.3. The effects of institutional changes on agricultural enterprises' productivity

When we analyse the effect of institutions on the performance of economic activities, the first problem we encounter is how we should define 'institution'. 'Institution' may refer to a number of different things and as of yet, there is no clear and unique definition of the word in academic literature (Daron Acemoglu, 2010). Institutions can be categorised into three types: (i) political institutions; (ii) economic institutions; (iii) cultural institutions or social institutions. Political institutions refer to government bodies or agents, while economic institutions refer to firms or groups of economic agents such as households, traders, or brokers. Cultural institutions or social institutions refer to other social groups in the society such as associations of gardeners or a football associations. This part is focused only on political and economic institutions, as they are the two major types of institutions and have the most significant effect on an enterprise's productivity.

Political and economic institutions can encourage or discourage agricultural firms to improve their productivity via unintended policies. Appropriate policies will encourage the development of firms, while biased policies will encourage some firms and discourage others. One important role of political institutions is to smooth the progress of contracting between financial institutions and borrowers or between different enterprises. Thus, if laws, courts and regulations protect the contracts, contracting will be possible. On the other hand, if laws, courts, and regulations promote bad behaviour, such as corruption, the competition between firms will be distorted.

Institutional changes refer to the transformation of a particular government system or the improvement of a firm. The relationship between institutional changes and agricultural firms' productivity has two parts, namely inside and outside. Outside effects refer to the influence of public policies on the performance of the agricultural firm. Inside effects refer to the improvement of working conditions or incentive policies on behalf of the board of directors of the firms. Certainly, the outside effects are easier to observe and measure, while inside effects are more difficult to collect data on, because internal information is sometimes private and not published. It is hard to judge the inside effects as more important than outside effects, or vice versa. Both effects can be either negative or positive, due to the quality of the policies made by responsible bodies.

7.4. Methodology

7.4.1. The model of agricultural households

For any agricultural household, it is assumed that they try to maximise their utility function as follows:

$$U^{Indi} = U^{Indi}(house\_product, firm\_product, leisure, socialized, house\_characters) \quad (7.1)$$

Utility is maximised subject to the household's budget constraints:

$$c = p * house\_product + w * labour\_supply + r * socialized \quad (7.2)$$

The problem here is that agricultural households may or may not know about the third term in the right hand side of equation (7.2). If they know, they might not know how to calculate this benefit. Normally, the benefit from investing in relationships is unobservable; however, we cannot deny that maintaining the relationship actually gives the owner the chance of gaining more value.

Agricultural households also face time constraints in that they cannot spend more than 24 hours a day working (on-farm production or off-farm for employment), entertainment or maintaining the relationships. Therefore, we have

$$Total\ time = House\_Working\_time + Labor\_time + Socialized\ time \quad (7.3)$$

They also face production constraints due to their available resources, which are related to technology, means of producing and other factors such as cultural and social norms.

$$Q = Q(House\_Working\_time, Land, socialized, other\_factors) \quad (7.4)$$

Here, it is assumed that: (i) agricultural households' productions are riskless; (ii) family labour and hired labour are perfect substitution and will be able to add directly; (iii) the households produce only one harvest for the whole year; (iv) agricultural households are price-takers in three markets (purchased-commodity market, staple market, and labour market);

(v) utility function of a particular individual satisfies all neo-classical conditions.[45]

The utility maximisation of agricultural households can be summarised as follows:

$$\max_{L,R} U(pQ(L, Land, R, A) + r*R, T - L - R; house\_characters) \quad (7.5)$$

subject to: $0 \leq R \leq T - L$

with: L being hours spent for doing house stuffs, W is working hours in the labour market, R is *socialised* in equation (7.1) and (7.4), r is rate of return of investment in social relationship and W is working hours of the labour market. Applying the Lagrangian function for optimal program yields:

$$\mathcal{L} = pQ(L, Land, R, A) + r * R + r*(T - L - R)$$

The first-order condition for L states that

$$r^* = pQ_L(L, Land, R, A) \quad (7.6)$$

where $Q_L(L, Land, R, A)$ is the derivative with respect to L.

The first-order conditions for R state that

---

[45] 1. **Complete.** For all x and y in X, either $x \succsim y$ or $y \succsim x$ or both
2. **Reflexive.** For all x in X, $x \succsim x$
3. **Transitive.** For all x, y and z in X, if $x \succsim y$ and $y \succsim z$, then $x \succsim z$
4. **Continuity.** For all y in X, the sets $\{x : x \succsim y\}$ and $\{x : x \precsim y\}$ are closed sets. It follows that $\{x : x \succ y\}$ and $\{x : x \prec y\}$ are open sets.

$$r = r^* - pQ_R(L, Land, R, A) \quad (7.7)$$

with $Q_R(L, Land, R, A)$ being the derivative with respect to R. Thus,[46]

i. $r = r^*$ if $R = 0$
ii. $r \neq r^*$ if $0 < R \leq T - L$

Dividing both sides of the equation (7.6) to r^* we have

$$\frac{r}{r^*} = \frac{r^* - pQ_R(L, Land, R, A)}{r^*} = 1 - \frac{Q_R(L, Land, R, A)}{Q_L(L, Land, R, A)} = 1 - \frac{MPR}{MPL} \quad (7.8)$$

where *MPR* is marginal product of *socialized*, and *MPL* is a marginal product of *L*. If we call r the rate of return of investment on social relationship, r^* can be seen as the opportunity cost of spending time on maintaining relationships rather than doing other tasks. In equation (7.8), *MPR* and *MPL* can be calculated by estimating the equation (7.4). It is assumed that agricultural households' production function is in Cobb-Douglass form, so

$$Q_{i,t} = A(K_{i,t})^\alpha (L_{i,t})^\beta e^{rR_{i,t} + \gamma Z_i + \varepsilon_{i,t}} \quad (7.9)$$

where
- $Q_{i,t}$ is output of household i at the time t
- $K_{i,t}$ is the crop land area of household i at the time t
- $L_{i,t}$ is the working hours of household i at the time t
- $R_{i,t}$ is social capital of household i at the time t

---

[46] These first-order conditions are derived from the Lagrangian function.

- Z_i is household's characteristic i
- A is the technology level of agricultural households and assumed to be a constant
- α and β are coefficients, $0 \leq \alpha, \beta \leq 1$.
- r is the rate of return of social capital investment
- $\varepsilon_{i,t}$ is the error term

Taking a natural logarithm for both sides of equation (7.9), we have

$$\ln(Q_{i,t}) = \ln(A) + \alpha \ln(K_{i,t}) + \beta \ln(L_{i,t}) + rR_{i,t} + \gamma Z_i + \varepsilon_{i,t} \quad (7.10)$$

Equation (7.10) will be estimated and reported in the empirical section.

7.4.2. *The model of agricultural firms*

It is assumed that agricultural firms have a production function in Cobb-Douglas form as follows:

$$Y_{i,t} = AK_{i,t}^{\alpha} L_{i,t}^{\beta} \quad (7.11)$$

where, $Y_{i,t}$ is the output of agricultural firms[47] at province i, time t. $K_{i,t}$ is the total fixed capital of agricultural firms at province i, time t. $L_{i,t}$ is the total labourers working in agricultural firms at province i, time t. α and β are coefficients. A represents the technology level of agricultural firms and is a constant. Additionally, equation (7.11) could be used as baseline model.

---

[47] Here, we assume that all agricultural firms have the same level of technology.

Taking log for both sides of equation (7.11) and adding the error term we have the following econometric model:

$$\log(Y_{i,t}) = \log(A) + \alpha \log(K_{i,t}) + \beta \log(L_{i,t}) + \epsilon_{i,t} \quad (7.12)$$

In equation (7.12) if: (i) $\alpha + \beta = 1$, constant return to scale; (ii) $\alpha + \beta > 1$, increasing return to scale; and (iii) $\alpha + \beta < 1$, decreasing return to scale.

To measure the effect of social capital presented by some dimensions of government measurements, we add some variables to equation (7.11). We then have

$$Y_{i,t} = AK_{i,t}^{\alpha} L_{i,t}^{\beta} e^{\lambda_1 hd\_policy_{i,t} + \lambda_2 trans_{i,t} + \lambda_3 pro_{i,t} + \lambda_4 ssb_{i,t} + \lambda_5 informa_{i,t} + \lambda_6 institution_{i,t}} \quad (7.13)$$

where
- $Y_{i,t}$ is the total revenue from agricultural business firms in province i at time t;
- K_{it} is the total fixed assets and long term investments of agricultural firms in province i at time t;
- L_{it} is the total labour at the time of interviewing agricultural firms in province i at time t;
- A is the technical level of agricultural firms and it is assumed that A is constant
- hd_policy is an assessment of firms regarding local government policy on human development in province i at time t
- $trans$ is an assessment of firms on transparency of local government in province i at time t

- *pro* is an assessment of firms on the proactive nature of local leadership in province *i* at time *t*
- *ssb* is a measure of the bias toward state-owned enterprises, equitized firms and other provincial companies by local authorities in terms of incentives, policy and access to capital in province *i* at time *t*
- *informa* is opinions of firms on informal charges in province *i* at time *t*
- *institution* is an assessment of legal institutions in province *i* at time *t*

Taking natural log of both sides of equation (7.13) and adding an error term we have an econometric model:

$$\ln(Y_{i,t}) = \ln(A) + \alpha \ln(K_{i,t}) + \beta \ln(L_{i,t}) + \lambda_1 hd\_policy_{i,t} + \lambda_2 trans_{i,t} + \lambda_3 pro_{i,t} + \lambda_4 ssb_{i,t} \\ + \lambda_5 informa_{i,t} + \lambda_6 institution_{i,t} + \varepsilon_{i,t} \quad (7.14)$$

Equation (7.14) will be estimated and reported on in the empirical section.

7.4.3. Datasets and variables

In order to estimate equations (7.10) and (7.14) we employed two datasets: (i) the Vietnamese Households Living Standard Survey (VHLSS) 2002, 2004, 2006, and 2008 for estimating equation (7.10); (ii) constructed panel data at provincial level from 2005 to 2009 by using several statistical texts.

Agricultural household model

The VHLSS 2002, 2004, 2006, and 2008 was conducted by the Vietnamese General Statistical Office (GSO) with technical support by the World Bank and the UNDP in Vietnam. The goal of the VHLSS is to provide information for calculating household incomes, living standards and the consumer price index. The VHLSS is a rich dataset with a wide range of variables ranging from economic to social indicators.

Firstly, all agricultural households were defined from the surveys by using their ID and employment status, combined with their income source. Additionally, the outputs of agricultural households were valued and deflated by regional and monthly consumer price indexes, which are provided by the GSO. Inputs and outputs of Vietnamese agricultural households were directly calculated using the survey questionnaires.

Crop land is measured in square meters, while working time is measured in hours. Other characters of agricultural households are (i) the gender of the head of households; (ii) education of the head of households and the head's wife/husband's education attainment; (iii) children's education attainment;[48] (iv) the characteristics of accommodation; (v) others.

Secondly, for measuring social capital at individual and family levels, some variables[49] in the VHLSS were employed. *Outdoor eating* expenditure was the first indicator for estimating social capital at the

---

[48] This characteristic is an important criterion; Coleman (1988) asserts that social capital of a family is the relationship between parents and their children. One of the most crucial connections among them is the school attainment of the children.

[49] The detailed statistical summary of these variables appear in appendix D.

micro level. It is suggested that Vietnamese farmers are maintaining their relationships with their relatives and friends by providing meals for outdoor meetings. At the meetings, one person may pay for the meal if he/she wants to share a personal celebratory event or celebrate a special event; alternatively, all participants will pay an equal amount of money for the meal. *Food expenditure* during holidays or spent on celebrations is the second indicator. Holidays and celebrations are events that all members of a family participate in together, perhaps travelling somewhere or celebrating a special family event. At that point in time, the relationships between parents and their children will be amplified; thereby, the role of the creation of human capital from social capital is implemented. *Pocket money for children* is the third indicator. The purpose of giving money to children varies among households and we cannot conclude the activity as either good or bad, due to the lack of information. *Contribution to parent funds and class funds* is the fourth indicator. Parent funds and class funds are used to pay for the outdoor activities of children who are studying at school and for buying teachers a gift on teacher days or to celebrate the New Year. All funds are voluntary.

Agricultural firm model

The results of an enterprise survey conducted by the GSO over a number of years have been employed to obtain data on capital and labour of agricultural firms at the provincial level. The survey began in 2001 and since then has been conducted regularly.

In the enterprise survey we can obtain data for revenue, fixed investment and the labour of agricultural firms at the province level.

There are three types of core data for estimating the baseline equations of production function for agricultural firms. By constructing panel data for a firm at province level we can take advantage of panel data analysis, and the other advantage is we do not have to worry about degree of freedom. Moreover, the effect of geography on production can be estimated. Another geographical variable is the distance between each province to Dong Ha City and Quang Tri Province (17th parallel). The distance is measured in kilometres.

The data on institution is employed from results of a survey series conducted by Dr Edmund Malesky and funded by USAID and the Vietnam Chamber of Commerce and Industry from 2005 to 2009. Using data from the survey, the Vietnam Provincial Competitiveness Index (hereafter PCI) was calculated. The PCI represents the opinions of thousands of enterprises doing business in Vietnam on economic governance and the regulatory environment in Vietnam. To date, in Vietnam, the PCI report has been used by just over 40 (out of 64) provincial People Committees in shaping their socio-economic development plan (SEDP) and improving the performance of governance activities. Therefore, the data from the PCI report can be seen as a reliable data source for the analysis of this chapter.

The PCI was calculated by using nine sub-indexes: (i) entry cost; (ii) land access and security of tenure; (iii) transparency and access to information; (iv) time cost of regulatory compliance; (v) informal charges; (vi) proactivity of provincial leadership; (vii) business support services; (viii) labour and training; (ix) legal institutions. Only sub-indexes iii, v, vi, vii, viii, and ix were employed to analyse the effect of institution performance on agricultural firms' productivity.[50] Detailed statistical information can be found in appendix D.

---

[50] '3. *Transparency and Access to Information*: a measure of whether firms have access to the proper planning and legal documents necessary to run their businesses, whether those documents are equitably available, whether new policies and laws are communicated to firms and predictably implemented, and the business utility of the provincial webpage.
   5. *Informal Charges*: a measure of how much firms pay in informal charges, how much of an obstacle these extra fees pose for business operations, whether payment of these extra fees results in expected results or "services," and whether provincial officials use compliance with local regulations to extract rents.
   6. *State-owned enterprise bias and competitive environment*: a measure focusing on the perceived bias of provincial governments toward state-owned enterprises, equitized firms, and other provincial champions in terms of incentives, policy, and access to capital.
   7. *Proactivity of Provincial Leadership*: a measure of the overall attitude of provincial officials as well as their creativity and cleverness in implementing central policy, designing their own initiatives for private sector development, and working within sometimes unclear national regulatory frameworks to assist and interpret in favor of local private firms.
   8. *Labor and Training*: a measure of the efforts by provincial authorities to promote vocational training and skills development for local industries and to assist in the placement of local labour.
   9. *Legal Institutions*: a measure of the private sector's confidence in provincial legal institutions; whether firms regard provincial legal institutions as an effective vehicle for dispute resolution or as an avenue for lodging appeals against corrupt official behavior' (Malesky, 2009, p. 10).

7.5. Empirical results

7.5.1. The estimated results for the agricultural household model

7.5.1.1. Social capital had a positive effect on agricultural households' income in 2002

Table 7.1. The effect of social capital on agricultural households' income in 2002

Dependent variable is natural log of households' income	
Variables	Coefficient
Natural log of households' land	0.819***
	(0.007)
Natural log of rest hour	0.486***
	(0.137)
Natural log of working hour	0.048
	(0.031)
Natural log of pocket money for children	−0.0002
	(0.004)
Natural log of fund to class	0.011***
	(0.004)
Natural log of holiday outdoor eating	0.033***
	(0.004)
Car (=1 if distance to car road <3 km)	0.064***
	(0.018)
Trans (=1 if distance to transportation < 10km)	0.039***
	(0.012)
Centre (=1 if distance to district centre < 10 km)	−0.021*
	(0.012)
Post (=1 if distance to post office < 5 km)	0.035**
	(0.014)
Market (=1 if distance to daily market < 5 km)	0.127***
	(0.016)
Regular (=1 if distance to periodic market < 5 km)	0.013
	(0.012)

Dependent variable is natural log of households' income	
Variables	Coefficient
Constant	−2.621**
	(1.324)
R^2	0.6345

Note: ***, ** and * are statistical significant at 1%, 5% and 10%, respectively. Numbers in parentheses are robust standard errors. Sample is weighted.

Source: Author's estimation

As can be seen in Table 7.1, the natural endowment presented by land is a significant source of income for agricultural households, while the longer the hours farmers work in the field, the less income they earn. Land is a very important production factor of agricultural households; however, under restriction of agricultural technique or the way of doing agriculture, Vietnamese farmers are on the frontier of their own production curve. When they spend more time in the field, farmers cannot allocate longer time to other efficient jobs. In fact, in rural areas, farmers need to spend only a short period of time on agricultural duties due to the small areas of their fields. Consequently, these farmers should reduce their time working on crop land in order to have more time for doing other beneficial jobs.

Looking at the effect of social capital on agricultural households' incomes, we can see that the situation is somewhat different. *Firstly*, the link between two generations in a family, which is proxied by two variables (natural log of pocket money for children and natural log of fund to class), has both negatively and positively affected agricultural household incomes. The concerns of parents about their children's schooling and their way of using money reflect an understanding of one another between two generations in the same family. Generally, it is assumed that high-income families can contribute to class funds much easier than lower income families; however, in this case, contribution to class funds does not reflect the income inequality between families. All parents have the same responsibility to the class fund. Moreover, in a negative sense, the aspect of pocket money for children implies that the relationship between two generations decay when income increases; however, this coefficient is not statistically significant, which means we can disregard the effect of this variable on farmers' income. *Secondly,* the maintenance of friendship or other types of relationships between people, which is presented by expenditure of meals outside agricultural households, and the expenditure for food on holidays, had a positive effect on agricultural household incomes in 2002. It is stated that in the case of Vietnamese rural areas, the traditional custom of joining in meals for special events is always a good decision. This traditional custom is still very important in rural life. The empirical results reflect the fact that the maintenance of rural household relationship helps farmers in gaining more benefits. The wealth of agricultural households is increasing by spending more on food, on holidays, or on organising meals outside the home to which friends and related people is invited.

Most social infrastructure variables have a positive effect on agricultural households' income, with the exception of the distance from the commune to the district centre. There is a significant difference between the incomes of farmers who live in more favourable areas with better infrastructure. The highest effect is the dummy variable of distance from commune to the closest daily market. Farmers in areas where the distance to the closest daily market is shorter have higher incomes than those living further away. Agricultural households in areas where the distance to the daily market is less than 5 km have a 12.5 percent higher income than others.

Agricultural households in communes that are closer to the transportation system have better income than others. Being closer to a transportation system means that these households have the advantage in delivering their production to market, whole seller and retailer. Farmers living in the communes where distance to the road is lower than 3 km have 6.4 percent higher income than others who are living in the commune where the distance is higher than 3 km. In addition, the commune is closer to the transportation system than the other communes, presented by the dummy variable, which gets value 1 when the distance from commune to transportation system is lower than 10 km and provides an opportunity to its agricultural households to have 3.9 percent higher income than other farmers.

In short, investment in social capital presented by social infrastructure, social relationship outside families and inter-family social connection has positively affected agricultural households' income. This means that in rural areas of Vietnam social cohesion has an important role in the life of farmers and their way of earning money. However, based on the empirical evidence, some types of investment in social capital do not help farmers much in gaining more money. This type of investment even takes time of farmers and then they waste time on inefficient things.

7.5.1.2. Social capital has positive effect on agricultural households' income in 2004

Table 7.2. The effect of social capital on agricultural households' income in 2004

Dependent variable is natural log of households' income

Variables	Coefficient
Natural log of households' land	0.152***
	(0.018)
Natural log of rest hour	0.624***
	(0.231)
Natural log of working hour	0.321***
	(0.115)
Natural log of pocket money for children	0.109***
	(0.013)
Natural log of fund to class	0.042***
	(0.014)
Natural log of outdoor eating in holiday	0.232***
	(0.019)
Car (=1 if distance to car road <3 km)	−0.054
	(0.057)
Trans (=1 if distance to transportation < 10km)	−0.015
	(0.031)
Centre (=1 if distance to district centre < 10 km)	−0.065**
	(0.031)
Post (=1 if distance to post office < 5 km)	−0.057
	(0.034)
Market (=1 if distance to daily market < 5 km)	0.111***
	(0.035)
Regular (=1 if distance to periodic market < 5 km)	−0.005
	(0.034)
Constant	−0.098
	(2.21)
R^2	0.3830

Note: *** and ** are statistically significant at 1% and 5%, respectively. Numbers in parentheses are robust standard errors. Sample is weighted.

Source: Author's estimation

In 2004, the effect of wealth endowment, proxied by natural log of agricultural land space, to agricultural income decreased significantly, while the effect of socialised hour to farmers' income increased dramatically compared to the year 2002. The estimated effect of agricultural land on income reflects the situation that agricultural households undervalue their wealth endowment; in fact, 2004 saw a large movement of agricultural labour to other sectors such as manufacturing and construction. Additionally, the effects of wealth endowment and socialised hours are statistically significant at 1 percent and have economic meaning. The high value of natural log of the rest hour of agricultural households implies that socialised activity might have a crucial effect on farmers' income.

The effect of the link between generations in farmers to agricultural household incomes was transformed from a negative in 2002 to a positive in 2004 and the effect is statistically significant at 1 percent. This means that the effect of the link on farmers' income was not consistent between 2002 and 2004; however, the results imply that there is a connection between inter-family social capital and the income of these households. In 2002, the value of the link variable was close to 0 and not statistically significant. Thus far, we do not have much evidence of the relationship between generations the income of farmers; however, it is too soon to conclude that this type of social capital has inconsistently affected agricultural household incomes.

Additionally, the concern of parents for their children in schools, where the human and social capital of children have been shaped, has not only enhanced their income but also improved the link between them, their children and the schools. At the micro level of social capital, the connection between parents, children and schools plays an important role increasing children's human and social capital, as well as the social capital of families. In some instances, this connection facilitates knowledge to parents about their children's status in school. In addition, the maintenance of the connection will improve parents' dynamic for earning more money in order to provide their children with better access for achieving deeper knowledge, which will provide children with opportunities for a better future. However, contributing more money to schools does not necessarily equals more knowledge for children. In fact, the value of the coefficient means that increasing 1 percent of the contributions to school funds will lead to an increase of 0.04 percent to the dynamic of earning more money.

Socialised activities such as having meals with family members, relatives, or friends have positively affected the income of agricultural households. Through activities such as these, the ties between individuals will be enhanced, thereby improving the chances of earning better incomes or finding better methods for agricultural activities, such as the appropriate way for using chemicals to treat their harvests. Accordingly, the opportunity for having more jobs during leisure time following the harvest will be increased.

Furthermore, social infrastructure did not affect the income of agricultural households in 2004 as it did in 2002. *Firstly*, in Table 7.2 the signs of social infrastructure variables can be seen; these include *car*, *trans*, *post*, and *regular*. These are all negative and not statistically significant. *Secondly*, only agricultural households who live close to the market will have better incomes, as opposed to other households who live far from the market.

7.5.1.3. The effect of social capital on agricultural household incomes in 2006

Table 7.3. The effect of social capital on farmers' incomes in 2006

Dependent variable is natural log of households' income

Variables	Coefficient
Natural log of households' land	0.361***
	(0.024)
Natural log of rest hour	1.194*
	(0.723)
Natural log of working hour	1.047**
	(0.508)
Natural log of pocket money for children	0.117***
	(0.024)
Natural log of fund to class	−0.010
	(0.025)
Natural log of outdoor eating in holiday	0.212***
	(0.045)
Car (=1 if distance to car road <3 km)	0.281***
	(0.093)
Trans (=1 if distance to transportation < 10km)	−0.104**
	(0.050)
Centre (=1 if distance to district centre < 10 km)	0.130**
	(0.055)
Post (=1 if distance to post office < 5 km)	−0.091
	(0.057)
Market (=1 if distance to daily market < 5 km)	0.087
	(0.055)
Regular (=1 if distance to periodic market < 5 km)	0.197***
	(0.051)
Constant	−0.478
	(2.518)
R^2	0.2802

Note: ***, ** and * are statistically significant at 1%, 5% and 10% respectively. Numbers in parentheses are robust standard errors. Sample is weighted.

Source: Author's estimation

As can be seen from Table 7.3, the endowment wealth of farmers, which is proxied by land, has strongly affected their income; the coefficient value is 0.36 and is statistically significant at 1 percent. The value of the land coefficient is lower than other countries in comparison with the value of land in the Cobb-Douglas production function of agricultural sector; for example, this coefficient is 0.46 and 0.43 for Indonesia and the Philippines, respectively (Y. Mundlak, D. F. Larson, & R. Butzer, 2002).

Additionally, the elasticity of the time for maintaining relations[51] in year 2006 was higher than years 2002 and 2004. The extreme value of the coefficient for bonding time during 2006 reflects the high appreciation of farmers of their relationships. They value the time they have for meeting relatives, friends, and outdoor eating. In rural areas of Vietnam, because of the situation that agricultural land is fragmented, time for meeting and maintaining relationships between family and friends is very important, as these meetings act as opportunities for farmers to find more opportunities to earn more income. In different areas, farmers regularly attend the village meetings, where they address their needs and hear the demands of other farmers. These meetings provide opportunities to access financial resources or provide information on how to access these resources. Because of village and commune meetings, many important issues that have previously been discussed by villagers will be addressed and needs may be met. Thus, the number of farmers who often participate in the meetings and the extreme value of the bonding time attributed to this during 2006 could explain how

---

[51] The database that I employed for this chapter does not have any information about what type of activities people use for the rest time. Therefore, here it is assumed that after working in the fields, farmers spend all their time on external and internal family relationships.

the knowledge of farmers about maintaining relationships and gaining information about common events are maintained.

The aspect of working time in year 2006 was still positive, as in years 2002 and 2004; additionally, its value was much higher than in the two previous years. The positive sign of this coefficient implies that the harder farmers work, the higher incomes they earn and this conclusion holds for three datasets. Under the assumption that agricultural technology and the way of cropping are the same between farmers, if agricultural households spend more time in their fields they will have a better chance to gain a higher income, especially in favourable weather situations. Uncomfortable weather will damage crop productivity; however, if farmers focus on their crops and take good care of them, they will have better solutions to minimize their crop's damaged due to bad weather than farmers who spend less time working on their crops.

Next, we will address the question of whether farmers can increase their working and socialising time simultaneously. When they increase their working time, farmers have to reduce their socialising time. Hence, farmers face a trade-off: working more or meeting more. It has been suggested that farmers do not know what type of activities are better, working or meeting; however, based on their specific conditions, they can decide what type of activities will be selected. Some farmers will choose to work more while others will choose to bond more. It depends on the farmer's individual goal and their calculation of the chance to gain more income. In short, farmers, and only farmers, can know and decide exactly what their needs are. This point is a very important signal for policymakers.

The link between parents and their children does not signify a perfect outcome. The coefficient of giving pocket money to children is positive and statistically significant at 1 percent, while the coefficient of contribution to class is negative and statistically insignificant. It is stated that farmers usually do not want their children to have a difficult life; parents will therefore save resources for the next generation. However, in this instance, the empirical results do not support the idea that social connections between households and schools will increase the income of agricultural households.

Once again, the meeting time, which is presented by the natural log of expenditure on outdoor eating in the holidays, has great influence. The high value of this variable implies that farmers highly assess the relationship between themselves and their relatives, and their friends. Eating outdoors on holidays may include expenditure on food for family members when they travel. Therefore, this variable can be explained in two ways; one is external while the other concerns internal relationships. For internal relationships, holidays are the time in which family members can meet each other to gain information and learn about each other's lives. For external relationships, holidays are the time farmers can spend more time with their friends, following a busy harvest. The positive sign of this variable implies that the increase of time for meeting between family members or friends will create for individuals a good chance to gain more money.

The availability of social infrastructure has significantly affected agricultural households' opportunities for doing business; however, the sign of some variables do not have any economic meaning, while others do not have statistical significance. Farmers who live in areas that have a better social infrastructure will have better opportunities for earning more money. In communes that are close to roads, farmers tend to have a higher income (28 percent) than farmers who live further away from roads.

Table 7.4. The summary of effect of social capital on agricultural household incomes in 2002, 2004, and 2006

Dependent variable is natural log of households' income

Variables	Coefficient		
	2002	2004	2006
1. Natural log of rest hour	↑***	↑***	↑***
2. Natural log of pocket money for children	↓	↑***	↑***
3. Natural log of fund to class	↑***	↑***	↓
4. Natural log of outdoor eating in holiday	↑***	↑***	↑***
5. Car (=1 if distance to car road <3 km)	↑***	↓	↑***
6. Trans (=1 if distance to transportation < 10km)	↑***	↓	↓***
7. Centre (=1 if distance to district centre < 10km)	↓**	↓**	↑**
8. Post (=1 if distance to post office < 5 km)	↑**	↓	↓
9. Market (=1 if distance to daily market < 5 km)	↑***	↑***	↑
10. Regular (=1 if distance to periodic market < 5 km)	↑	↓	↑***

Note: ***, ** and * *are statistically significant at 1%, 5% and 10% respectively.*

Source: Author's estimation

As can be seen from Table 7.4, only variables (1), (4), and (9) hold identical signs of effect of social capital on agricultural household incomes, while the rest have a different sign. The up arrows mean that when the variable is increased, agricultural household incomes will be raised. The down arrows mean that when the variable is increased, agricultural household earnings will decrease. This situation shows us the complex effect of the dimensions of social capital on the income of farmers at the household level.

7.5.2. The estimated results for the agricultural firm model

Initially, the baseline model with only three variables—firms' revenue, labour, and capital—was estimated by using generalised linear models for testing the type of returning to scale of the firms. For instance, if the model has $\alpha + \beta$ larger than 1, we could conclude that the firms were increasing return to scale. It is expected that at the provincial level, agricultural firms are decreasing return to scale due to the industrialisation policy of the Vietnamese government. Using a generalised linear model in estimating equations (which are presented above) led to the results in the following tables.

Table 7.5. Result of estimating the baseline model

Dependent variable is natural logarithm of firm's revenue at provincial level

Variables	Model	
	Full sample (from 2000 to 2008)	Sub-sample (from 2005 to 2008)
Log(K)	0.243***	0.181***
	(0.036)	(0.057)
Log(L)	0.736***	0.837***
	(0.042)	(0.065)
Constant	−2.242***	−2.313***
	(0.194)	(0.286)
Log likelihood	−699.403	−313.092
N	548	245

Note: *** is statistically significant at 1%. Numbers in parentheses are Observed Information Matrix (OIM) standard error.
Source: Author's estimation

As can be seen from the above table, the total of K and L coefficients closes to 1. The total of coefficients of K and L in the full sample is 0.98, while the sum in the sub-sample is 1.02. This means that agricultural firms at the provincial level in Vietnam are facing the constant return to scale. Additionally, the higher value of labour's coefficient in both models reflect the real situation, i.e., that agricultural firms in Vietnam are labour-intensive.

Next, it is suggested that the weather may have significantly affected the productivity of the crop and as a result, the operation of agricultural firms. In order to estimate the effect of the weather, I added one more variable, which is the natural logarithm of distance between the centre of each province to the 17th Parallel in Quang Tri Province plus 1 into the estimated model above. The result is reported in the table below:

Table 7.6. Estimated the effect of weather on agricultural firms

Dependent variable is natural logarithm of agricultural firm revenue at provincial level.

Variables	Models	
	Full sample (from 2000 to 2008)	Sub-sample (from 2005 to 2008)
Natural logarithm of Capital	0.290***	0.220***
	(0.036)	(0.059)
Natural logarithm of Labour	0.701***	0.803***
	(0.041)	(0.066)
Natural logarithm of (distance + 1)	0.192***	0.141*
	(0.035)	(0.054)
Constant	−3.469***	−3.199***
	(0.296)	(0.444)
Log likelihood	−685.157	−309.749
N	548	245

Note: ***, * are statistically significant at 1% and 10% respectively.
Source: Author's estimation

Obviously, the provinces where there are longer distances from Quang Tri control better quality natural resources in terms of agricultural harvesting. The value of the coefficient present for weather pattern,

Log(distance+1), is positive and statistically significant at 1 percent. Farther provinces to the North and South actually have better quality land, and as such, their agricultural firms have higher productivity than the firms in Quang Tri. Note that the weather and land quality in the middle of Vietnam are not in favour of agriculture. As can be seen in Table 7.6, under the effect of a weather proxy, the values of labour and capital variables are close together; however, their total is less than 1, which implies the decreasing rate of return to scale of agricultural firms at provincial level.

To capture the effect of social capital that is presented by a set of variables, equation 7.14 has been estimated. The detailed result is reported in Table 7.7 below.

Table 7.7. The effect of social capital on productivity of agricultural firms at provincial level from 2005 to 2008

Dependent variable is natural logarithm of agricultural firm's revenue at provincial level.

Independent Variables	Models		
	I	II	III
Natural logarithm of Capital	0.200***	0.229***	0.239***
	(0.061)	(0.061)	(0.061)
Natural logarithm of Labour	0.791***	0.757***	0.744***
	(0.069)	(0.067)	(0.068)
Human capital Policy		−0.081**	−0.085**
		(0.036)	(0.036)
Legal institutions	0.133***		0.072
	(0.048)		(0.052)
Transparency and access to information		−0.071***	−0.062**
		(0.027)	(0.027)
Informal Charges	0.075		0.055
	(0.075)		(0.079)
SOE Bias and Competition Environment		−0.142*	−0.159**
		(0.074)	(0.076)
Proactivity		0.206***	0.180***
		(0.045)	(0.048)

Dependent variable is natural logarithm of agricultural firm's revenue at provincial level.

Independent Variables	Models		
	I	II	III
Constant	−3.071***	−1.433**	−1.80**
	(0.578)	(0.603)	(0.709)
Log likelihood		−281.560	−270.951
N		224	224

Note: ***, ** and * are statistically significant at 1%, 5% and 10% respectively. Numbers in parentheses are OIM standard error.
Source: Author's estimation

Overall, the effects of proxies of social capital on agricultural firms' revenue at the provincial level are complicated and not easy to interpret. Their complicated nature is due to the complexity of the term 'social capital'. Social capital is a multifaceted economic term and we cannot encapsulate all dimensions of social capital into one variable. Therefore, the empirical results of estimating the effects of social capital on agricultural firms should be presented in the matrices of effects of several items of social capital.

In the three strategies, the results reflect the fact that all agricultural firms in Vietnam are labour intensive. The values of the coefficients of labour are significantly higher than the values of capital coefficients. The contribution of labour to agricultural firms' revenue is in a range varying from 0.74 to 0.79, while the fraction of capital is from 0.20 to 0.24. The negative value of the coefficient that represents human capital policy shows us that the current policy on training labour is decreasing firms' revenues. This result is consistent with the result of the Provincial Competitive Index survey conducted by USAID, where firms complained about the quality of labour (Malesky, 2009, 2010).

The reform of legal institutions means that a stable and predictable set of implementable laws will be available for firms as an input for making decisions to undertake costly investments. The high value of the legal coefficient shows us the assessment of agricultural firms concerning legal institutions at the provincial level. However, legal institutions in model I are statistically significant at 1 percent and positive, while in model III, this variable is not significant. This implies that legal institution reform has positively affected the revenue of agricultural firms at the provincial level; however, the empirical results are not stable. Clearly, firms with confidence in predictable laws will invest more in the sector, leading to their investments enhancing their revenue. Based on the analysis of the World Bank using the PCI survey database, the team members who wrote the Vietnam Development Report conclude that 'provinces where firms feel more confident in the predictability of laws also have more firms planning to invest' (Worldbank, 2009, p. 82).

Transparency and access to information constrain the growth of revenue of agricultural firms, because the current information system continues to favour the administrative report system. Consequently, firms may not find worthwhile information related to their business on provincial websites. In Vietnam, agricultural firms' plans still rely on the master plan of provincial and central government. The way of planning at local and central levels significantly influences the opportunities of firms. The negative value of the coefficient of transparency and access to information reflects the current situation in Vietnam and is consistent with the report outcomes of some international non-governmental organisations, such as the World Bank and USAID. Lack of access to information regarding planning at provincial and central levels of agricultural firms has limited the room for development of these firms, not only in the short term, but also in the long term (Worldbank, 2009). Moreover, the negative value of the coefficients of models II and III also reflects the negative impact of decentralisation inside Vietnamese government bodies, which is reported in the World Bank reports.

Informal charges have positively affected agricultural firms; however, the value of this coefficient is not statistically significant. The result implies that agricultural firms need to lobby government officers in order to have favourable supports. The reliance of firms on government officers, such as in Vietnam at present, is a characteristic of a low-developed market economy where government staffs have too much power and less democratic procedures or participation of individuals. Mass organisation and anti-corruption mechanisms do not work efficiently in this context. However, it is too soon to conclude that informal charges significantly affects the performance of agricultural firms at the provincial level, as more research and data is needed on this topic.

Additionally, the current strong support for state-owned enterprises policy of both provincial and central government does not offer much help for the improvement of agricultural firms at the provincial level. The statistical significance at 10 percent and 5 percent in the estimated models I and II respectively, indicates that the revenue of agricultural firms has decreased from 2005 to 2008 by implementing the policy. Agricultural firms should have strong support from government due to the speciality of agricultural business. However, in the situation of Vietnam and in the given time frame, support from local and central governments is inefficient. Moreover, the competitive environment for doing business is strongly influenced by policymakers. Biased support to SOEs has damaged the competition environment, because private companies cannot access credit sources or approach particular projects. The negative value of the coefficient can be understood when considering that private companies face biased policies from both local and central government.

So far, the macroeconomic reform, which was implemented in 1986, has led to the improvement of local and central government bodies from one year to the next. The progress has been captured by a series of studies and reports conducted by international nongovernmental organisations such as the World Bank and the United Nations Development Programme and USAID. The empirical results of this chapter give a reform outcome that is macroeconomic in nature, and is well captured by employing proactive variables from the Provincial Competitiveness Index survey. The statistical significance of proactive variables and the high value of the coefficient in models II and III (Table 7.7) implies that a proactive approach on the part of local governments encourages agricultural firms to increase their revenues.

Table 7.8. Estimated results of the time-varying decay inefficiency model

Dependent variable is natural logarithm of agricultural firm's revenue.

Variables	Coefficients
Natural logarithm of Capital	0.224***
	(0.037)
Natural logarithm of Labour	0.689***
	(0.040)
Distance	0.00091***
	(0.00016)
Year	0.166***
	(0.026)
Constant	−333.831***
	(52.722)

Dependent variable is natural logarithm of agricultural firm's revenue.	
Variables	**Coefficients**
Wald $\chi^2(4)$	1068.39
μ	1.933***
	(0.563)
η	−0.023*
	(0.013)

Note: *** is statistically significant at 1% respectively. Numbers in parentheses are standard error.
Source: Author's estimation

In the time-varying decay inefficient model, the effect of climate patterns, which is presented by a distance variable, is statistically significant at 1 percent, but its value is too small, very close to 0, while the coefficients of labour and capital confirm the idea that agricultural firms are faced with a constant return to scale by the total of $\alpha + \beta = 0.91$. The test of the linear hypothesis has resulted in Wald $\chi^2(4)$ with the value of 1068.39; as such, the hypothesis that the model is not linear has been rejected. The value of η is −0.023, smaller than 0, so the degree of inefficiency of agricultural firms increased over the time period from 2005 to 2008. Consequently, the level of inefficiency of agricultural firms at the provincial level increases to the base level. In order to understand the effect of institutions on the technical progress of agricultural firms at the provincial level, the function where technical inefficiency is a dependent variable has been estimated and is presented in the following table.

Table 7.9. Effects of social capital on technical inefficiency

Dependent variable is technical inefficiency of agricultural firms at provincial level.

Independent Variables	Models		
	I	II	III
Human capital Policy		−0.076***	−0.074***
		(0.012)	(0.012)
Legal institutions		0.058***	0.058***
		(0.021)	(0.022)
Transparency and access to information	0.040**		0.032*
	(0.020)		(0.017)
Openness	−0.003**		−0.004**
	(0.0018)		(0.002)
Informal Charges		0.100***	0.093***
		(0.025)	(0.026)
Equity	−0.041**		−0.029*
	(0.019)		(0.017)
SOE Bias and Competition Environment		−0.111***	−0.118***
		(0.027)	(0.027)
Proactivity		0.067***	0.068***
		(0.014)	(0.016)
Constant	0.025	−0.107	−0.011
	(0.039)	(0.225)	(0.240)
Log pseudo-likelihood	−73.795	−44.838	−42.988
N	224	224	224

Note: ***, ** and * are statistically significant at 1%, 5% and 10% respectively. Numbers in parentheses are robust standard error.
Source: Author's estimation

In model I, only indicators that are related to the transparency index of the PCI survey were used for estimating purpose. As can be seen in the table, the transparency and access to information has led to an increase of technical inefficiency of agricultural firms, while the other two dimensions of the transparency component of the PCI index, openness and equity, have decreased the technical inefficiency of agricultural enterprises at the provincial level.

In model II, other variables that are not related to the transparency index of the PCI survey were employed for estimating. Human capital policies of provinces are helping agricultural firms to decrease their technical inefficiency, while informal charges have led to the increase of technical inefficiency of agricultural enterprises at the provincial level. Interestingly, the current competitive business environment of Vietnam has brought about a decrease in the technical inefficiency of firms; however, legal institutions and proactive local and central authorities are instead contributing to increasing technical inefficiency in these firms.

In model III, which is reported in the last column of Table 7.8, the results do not differ much from those of models I and II. The only difference is a small change in the value of coefficients and their standard errors. Overall, the effect of social capital, which is presented by a number of institutional variables on technical inefficiency of agricultural firms, is reported in the table below.

Table 7.10. The summary of the effect of social capital on technical inefficiency of agricultural firms from 2005 to 2008 at provincial level

Human capital policy	↓	
Legal institutions		↑
Transparency and access to information		↑
Openness	↓	
Informal charges		↑
Equity	↓	
SOE bias and competition environment	↓	
Proactivity		↑

Source: Author's estimation

Table 7.10 summarises the effect of institutions on technical inefficiency. The up arrows imply the increased relationship between institutional variables and technical inefficiency, while the down arrows show the decreased connection between institutional variables and technical inefficiency. There are five factors that increase technical inefficiency when they are increasing, whereas there are four dimensions of institutional variables decreasing technical inefficiency when they are increasing.

7.6. Policy implications

Social capital has considerably affected the income of agricultural households. The cohesion of family members, social ties and social infrastructure can be seen as three major dimensions of social capital affecting agricultural household earnings. The effects of social capital on income of farmers are complex. Therefore, public policy encouraging farmers to invest in social capital is not easy to achieve.

Firstly, formal institutions should be developed and improved by the government in order to provide good quality services to people and reduce the technical inefficiency of firms. 'Formal institutions' refer to government bodies. In the case of Vietnam, formal institutions have considerably influenced economic activities through laws, legislative documents such as decisions, circulars or resolutions, and other rules.

'Institutions' encompasses different types, such as laws, social arrangements, regulations, enforcement of property rights and others. The policy implication on institutions should take into account all its types, but only formal institutions can be seen as the most effective type. Informal institutions such as mass organisation and social arrangement may have an effect on household incomes, but we do not have sufficient evidence regarding this relationship.

Secondly, time for working and socialising has positively affected agricultural household income levels. Thus, government should provide a supportive infrastructure for these activities such as community halls, entertainment areas, and other meeting places for citizens. Evidently, socialising hours have helped people to relax and maintain their relationships with not only their relatives but also their friends and have helped them make new friends. New and old relationships offer farmers opportunities to gain new jobs and opportunities to access financial resources. The sharing of information between members of a network will help farmers to solve problems they may encounter at any particular point.

However, because the length of a day is restricted, local governments should provide training courses on how to manage time. The training courses should be conducted yearly or quarterly, depending on the budget of localities. Time management skills are essential for people to lead a balanced life, which includes both working and socialising time. It is important to both individuals of the public and government officers that time management skills are seen as a vital skill for managing work and their lives in general. Time spent on work can generally not be used for socialising; therefore, individuals face a trade-off between working and socialising. In order to make suitable decisions, people need to be trained on how to efficiently use their time.

Thirdly, the concept of social capital provides a connection between families and society in terms of economics. Families are the foundation of society, while the economics of families is one of the most important agents of economic development in a country. Families contain social capital, as does each citizen of a country; the total social capital of a society can be viewed as an aggregation of citizens' social capital. Because social capital can change relationships among individuals and facilitate actions, it can change the economic decisions of persons or groups in order to gain more benefit.

Fourthly, to improve and develop rural infrastructure, and to change rural areas and promote socio-economic development, the Vietnamese government should invest 100 percent in five essential infrastructure programmes, including transport on the main roads of communes, schools, clinics, cultural houses, and Community People's Committee headquarters. The government should provide partial funding for other infrastructure works such as main roads of villages, water supply stations, commune stadiums, village cultural houses, village sport areas, rural electricity, markets, and rural communication systems.

7.7. Conclusion

This chapter examined the relationship between investment in social capital at micro and macro levels and agricultural households and firms. It employed three survey datasets to estimate relationships between social capital and farmers' income and two secondary datasets to estimate correlation among social capital and agricultural firms' revenues at provincial levels. Despite the fact that not all social capital variables are statistically significant, the empirical results are accurate and straightforward. The correlation between investment in social capital and farmers' income levels and agricultural firms' revenue has been caught by a set of social capital variables in order to present some dimensions of this type of capital. The empirical evidence shows that there is a strong correlation between social capital and agricultural firms' revenues at provincial levels, while the relationship between social capital and farmers' incomes is more complex and not always statistically significant. Therefore, policy implications have been carefully conducted and based only on positive and consistent outcomes from empirical evidence.

Chapter 8

PUBLIC POLICY FRAMEWORK FOR FURTHER AGRICULTURAL DEVELOPMENT

8.1. Introduction

This chapter will present a system of public policies that addresses human and social capital development for agricultural development in Vietnam. Human capital has two major elements: human health capital and human education capital, while 'social capital' is a more complicated term and contains more elements than human capital. The most significant element of social capital at the micro level is interfamily relationships between parents and their children. The second significant element of social capital concerns institutions. Agricultural development in Vietnam is driven by physical capital; therefore, agricultural development in Vietnam is horizontal. The need to develop a more productive agricultural system in Vietnam now requires not only the development of physical capital, but also the development of human and social capital.

The historic and current evidence concerning the important role of government in facilitating structural transformation in all successful countries may not be adequate for examining an idea that has long been discussed in economic literature. Many economists agree that government intervention is necessary for the successful transformation from an agricultural economy to an industrial economy. They usually claim that government is an indispensable ingredient for economic development. Additionally, successful transformation requires a general framework that can be employed as guidance for policymaking. As Charles Schultze, chairman of the Council of Economic Advisers for US President Jimmy Carter said,

> The first problem for the government in carrying out an industrial policy is that we actually know precious little about identifying, before the fact, a 'winning' industrial structure. There is not a set of economic criteria that determine what gives different countries pre-eminence in particular lines of business. Nor is it at all clear what the substantive criteria would be for deciding which older industries to protect or restructure. (Schultze, 1983, p. 5)

Promoting agricultural development in Vietnam therefore needs to employ a dual model in which agricultural development policy should be considered in a relationship with the industrial sector. The need for higher economic growth requires the rapid development of a more productive and skilled workforce. Meeting the demand for skilled and healthy workers in a rapidly growing and transforming economy is already proving to be a significant challenge. International experience suggests shortages of skilled and healthy workers will limit the growth rate of Vietnam's economy over the long term. As a large part of Vietnam's workforce works in agriculture, workers need to be trained or retrained at vocational training centres in order to meet the demands of the industrial sector.

The importance of social capital as a factor in economic development is now widely recognised. Researchers have to date focused on the macro element of social capital, i.e., institutions. Institutions are not buildings or organisations; rather, they are the rules that households, firms and government have to follow and interact with. In Vietnam Local level planning is now attracting the participation of citizens. Citizens are part of agricultural development as well. Legal services are provided to citizens and professional media circulate important events. These are among the root causes of agricultural development and the transforming of agriculture into a new stage in Vietnam.

The following parts of this chapter are organised as follows. Section 2 will present a set of public policies focused on human capital. Next, section 3 will present a set of public policies emphasising social capital. Regarding social capital, institutions and interfamily relationships will be two focal points. The final section is the conclusion.

8.2. Public policy implications on human capital development

8.2.1. Policy implications on education

Part of chapter 6 already presented policy implications on education. This chapter, however, provides a set of more comprehensive public policies on education as it relates to further agricultural development. For entering into the new stage of agricultural development, central and local governments should choose appropriate policies for doing so. Yet difficulties may arise in terms of what these policies are and how they operate comprehensively. Based on the mix of empirical results presented in chapters 5, 6, and 7, and with consideration to the policy debate presented in chapter 2, the set of public policies will be presented broadly here. It is suggested that the Vietnamese government should carefully consider the policy implications presented in this thesis. The reason for this is that the policies are built up from derived empirical results.

Some key challenges in education with a focus on vocational training

There is no doubt that improvements in Vietnam's education system have contributed to Vietnam's economic growth. It is also doubtless that higher education has benefited the younger population in finding satisfactory employment, leaving rural areas and escaping from agricultural work. On the other hand, for most Vietnamese farmers, for whom the highest education certificate is usually lower or upper secondary schooling, participating in a competitive labour market is still very difficult. Therefore, this part focuses on vocational training education, which is more pointedly related to farmers' current situation than to higher education. The reason for this is that a large part of current Vietnamese farmers were soldiers in the Independent War or were living in poor conditions; as such, they did not have many chances to explore further study options when they were younger. A significant part of these individuals remained in agriculture, while a smaller part left the agricultural and rural sector to attend university.[52]

In the past, vocational education has not had appropriate realisation in Vietnam, both by schools and state authorities. Currently, however, the attitudes of schools, learners and authorities toward vocational training are changing. Formal education has long been the only pathway into state careers.[53] The two main reasons for this are (i) low development of business; (ii) the poor quality of vocational training education. However, attitudes are changing among students. A large part of rural workers are now in vocational training schools. Between 1998 and 2008, enrolment in vocational training schools grew by four times and in 2009 stood at 1.7 million students. By 2010, nearly half of Vietnam's districts had at least one vocational training school. From 1997 to 2010, the rate of vocationally trained workers increased from lower than 10 percent to an estimated 30 percent in 2010.[54] Thousands of vocational training centres were set up

---

[52] Actually, Vietnam does not have a comprehensive survey regarding this statement. Therefore, researchers do not have much data to estimate the fraction of them among their labour classes.

[53] This situation was unlikely in developed countries, where education leading to careers in the state is one of many competing demands.

[54] http://www.molisa.gov.vn/news/detail/tabid/75/newsid/50067/laguage/vi-VN/Default.aspx?seo=Hoi-nghi-"Lay-y-kien-Doanh-nghiep-ve-Du-thao-Chien-luoc-phat-trien-day-nghe-giai-doan-2010-2020"

across the country; however, the country is only now raising systems of certification. Currently, every province is keen on establishing its own university, but less interested in vocational training schools; in particular, several industrial zones does not have any plans for establishing vocational training education.

Generally, vocational training education remains supply-driven and uses outdated training methods. Cooperation between schools and enterprises lags. The quality of teaching and teaching facilities of vocational training is poor due to insufficient qualified teaching staff and inadequate support from public authorities. Consequently, the needs of enterprises and developments in this segment of the labour market do not properly match.

Practically, vocational and professional training in Vietnam are currently managed by two ministries, the Ministry of Labor, Invalid, and Social Affairs (MOLISA) and the Ministry of Education and Training (MOET). This situation has resulted in two potential consequences. The first is that the effectiveness and strategic development of vocational training in Vietnam continues to be compromised. The second consequence is the confusion about the mission and value of vocational training. Vocational training is managed by MOLISA, whereas professional education schools and colleges have in the past been managed jointly by MOET and other specific functional ministers. Therefore, there is a desire to have all vocational and professional education schools managed by one ministry. Vocational and professional education students are also required to study certain elements of the standard upper-secondary school curriculum. The mixed model previously conducted in the past and referred to as vocational high schools is now being called on to be re-established. This model is similar to polytechnic schools in Singapore and offers both standard and vocational education, while also providing students with the possibility of further study at a four-year university level.

In short, in terms of vocational training education, access to more and higher quality technical and professional education is urgently needed. Principal challenges arise from the need to

- achieve rapid improvement in raising the number and quality of skilled workers;
- develop a vocational training education system in which excellence in teaching and learning environments is emphasised and encouraged;
- better and more strategically align education and training with current and projected demand for skilled labour;
- the fundamental challenge is to rapidly expand the quantity and quality of skilled labour. Doing so will require the creation of more and better training opportunities, the reduction of barriers to education and skills acquisition, improvements in quality and relevance of education, and addressing challenges in educational policy and administration.

In order to overcome these challenges, Vietnam must work to fix two features of the vocational training education system that are no longer appropriate for the country's needs. *First*, Vietnam needs to shift the locus of advanced research from functional ministries to universities. The old model followed the Soviet Union model, which is not a good fit for the current context in Vietnam. *Second*, the collaboration between MOLISA and MOET needs to be tightened. Under the management of MOET, vocational training education will be a more prominent feature of the mainstream education system.

Policy implications

The Vietnamese government's new focus on vocational education is sensible, but the content and modalities of the current training programmes need to be reconsidered carefully by different levels of authorities. Evidently, many programmes are not yielding their intended benefits. In Vietnam, vocational schools at the upper-secondary level are still in short supply. Studying the experiences of other countries, such as Singapore, Korea, and the Czech Republic, may help Vietnam to make vocational training education a more integrated feature of the mainstream education system.

Training programs targeting groups should not disregard agricultural workers, especially young farmers. Training that promotes employment in rural areas through diversification and entrepreneurship should be encouraged. This implies that the training schemes be based on an intimate knowledge of the local setting. To guide the development of vocational education, the government has constructed and issued a list of 301 careers for which vocational colleges are to focus training on and 385 careers for secondary vocational schools. However, in practice, the lists have been applied poorly at local levels. Learners have been trained in unnecessary knowledge that they cannot apply to their work or create a new job with.

There is a debate among government officials about the system of vocational training. On one side, many officials agree that businesses should set up training schools of their own. MOLISA has put forward supportive policies to the government for corporations that have opened training centres. Other officials suggest that vocational schools should sign a contract with businesses in order to know demand of firms before they select students. Both of the above approaches have been attacked by the other side. The opposed parties argue that these approaches create a situation in which learners will be provided with narrow training and as such, may encounter difficulties finding employment. Consequently, the opposed company suggests to the government that vocational training should instead be supplied as part-time study. Learners can learn at school during evenings and work at businesses during the day time. If instruction is full time, there is a particular need for making sure that training addresses the requirements of the labour market.

Vocational skills and knowledge are typically obtained from on-the-job experience (or 'learning-by-doing') and formal education. Therefore, the government should build up an effective system in which formal or informal apprenticeship and formal instruction are provided simultaneously to learners. Empirically, learners have many benefits from on-the-job training and company-based training schools; however, without formal instruction, learners do not have enough theoretical knowledge to improve their skills.

8.2.2. Policy implications on health

This part will present comprehensive policy implications on health as it relates to agricultural development in Vietnam. A population with better health is not only the goal of Vietnam, but also of other countries worldwide. Moreover, in the progress of agricultural development in

Vietnam, a better health rural population is the key element of the new rural development strategy of the Vietnamese government. Based on the empirical results, this part focuses on health policy implications related to the current development stage of the strategy.

Some key challenges facing the health system

Initially, there is a need to make a distinction between public health needs and the demand for health services. Public health needs refer to the expansion of quality health services to all segments of population and geographical regions and implementing means and measures for effectively responding to changing patterns of human health risks and shocks. Demand for health services refers to special challenges related to the design and conduct of health policy in a market economy. Along with the development process, health services emerge to the economy not only as an important service sector, but also as a crucial economic sector. Moreover, regarding health policy, a growing economy has many obvious benefits. Increases in public and private spending on health are an example of these benefits. The demand also requires improvement in the range and quality of health services conditions. However, in Vietnam, the difference between interests of health service providers and public health needs is expanded. Hence, the government faces a policy challenge. Harmonising the interests of health service providers and public health needs is not an easy task and requires significant financial and human resources.

Second, there is instability between essential health services and more sophisticated healthcare technologies. With an increase in Vietnam's wealth, there has been intensification in the consumption of health services and the demand for sophisticated healthcare. Ultrasound machines, which were hardly found in district hospitals in the 1990s, can now be found in most district hospitals and many private clinics. Computed tomography (CT) scanners and magnetic resonance imaging (MRI) machines are progressively more accessible. According to reports from the Vietnamese Ministry of Health (MOH), in 2008, nearly 180 million consultations, over 205 million lab tests and over 10.6 million ultrasound scans were completed. These figures are much higher than in 2000, when nearly 132 million consultations, 46.5 million lab tests and 2.6 million ultrasound scans were carried out.

In contrast, community health centre (hereafter CHC) infrastructures in poorer areas, especially mountainous regions, were identified as weak. For example, in Tay Nguyen, the availability of electricity for 33 percent of CHSs was either irregular or non-existent. Additionally, 20 percent could not access clean water and 33 percent lacked a waste dumping site (Tuong, 2007). Many CHCs did not have restrooms or bathroom facilities. A telephone was found installed in only 66 percent of CHCs in Dak Lak province and Kon Tum. Ninety percent of CHCs in Ha Giang lacked seven of the essential items needed to complete their sets of equipment. All CHCs had inadequate drug supplies. The drugs available in CHCs only met 20 percent of drug needs. Insufficient supply of essential drugs was evident in all these CHCs (Nhu, 2003).

The status of the health sector in Vietnam (as in other developing countries) as an income/profit generating sector carries risks. Interventions to promote healthy lifestyles (limiting smoking, alcohol abuse, e.g.) and diet are lacking, while the overuse of ultrasound machines in the country is increasing, especially on women during pregnancy.

Finally, people's right to health protection is still far from perfect. Although the Law on Protection of People's Health stipulates people's right to health protection, there has been no real argument on the implementation of such a right. To the vast majority of the Vietnamese population, there is little awareness of the perception that people are the right holders and government bodies are functional carriers. In Vietnam, health services have to date been considered a gift, a charitable action of the government; therefore, people are supplied with whatever hospitals have, otherwise it is considered a commodity that needs to be paid for in order to receive it.

Most popular newspapers in Vietnam have a section asking for donations to help people who require extreme health treatment. Moreover, groups of students are willing to do the same. Quite often, those who read Vietnamese newspapers regularly come across stories about children who have to stop their studies to go to work in order to earn money for paying for his/her parents' debt due to expensive treatment costs in hospitals. These situations usually happen to rural people or agricultural household members. Children in these situations do not always earn enough money and his/her parents need help in the form of donations from people. This fact has led to the public perceiving government as not reacting to people's unmet healthcare needs.

Policy priorities in the health sector

The first priority is that a well-functioning governance structure should be able to regulate, coordinate, and facilitate key players' participation in meeting common targets set for the system.

The second priority is educating the population and health professionals about the proper use of technology. Additionally, striking the right balance between the improvement of essential services and investment in high-tech equipment are among the key factors for receiving 'more health for less money'.

The third priority is increasing the awareness of the population, the government and policymakers about people's right to health, and the government's duty to protect people's health, build the capacity of the right-holder to demand their rights and for duty-bearers to fulfil their duty. This will require strong support from international partners, the broad participation of the public and the engagement of policymakers.

8.3. Public policy implications on social capital development

The importance of effective institutions as a factor of agricultural development is now widely recognised. The importance of institutional reform in Vietnam was well documented in the Vietnam Development Report 2010, called Modern Institutions, which was carried out by the World Bank. The report provides an extensive analysis of institutional matters; there is therefore no need to cover the same argument here again. Consequently, this part of the chapter emphasises some high-priority issues that should be addressed by the government and communities for developing agriculture and rural areas in the coming decade.

Some key challenges of public policy for social capital development

The first challenge is the efficient development of a legal framework related to agricultural development. There are two crucial elements for further development of the legal framework:

(i) The development of an increasingly sophisticated agricultural sector needs to be supported by a legal framework that maintains simplicity in such areas as property rights, land ownership and contractual obligations. An effective related legal framework will reduce business risks and transaction costs for farmers.

(ii) Respect for the law by citizens and government officials and the development of an approach that resorts to the law can assert citizens' rights and act as a check on misbehaviour by officials that will improve respect for the government and discourage recourse to other avenues that may influence officials (e.g., bribery).[55]

A good deal of work has been put into developing Land Law, which includes agricultural land. In recent years, many legal documents have been revised, amended, or newly issued to meet the demand of economic development and the requirements of membership to the WTO. Since Vietnam began its reformation, Law on Investment and Law on Enterprises were issued and have been adjusted regularly. These adjusted laws have changed Land Law and has directly affected farmers' living conditions. However, adjustment of legal frameworks need better insight and bigger

---

[55] For example, see the article in the *Financial Times*, 'Farmers fight highlights Vietnam's inequality', available at the following link: http://www.ft.com/intl/cms/s/0/8625ddc6-4352-11e1-8489-00144feab49a.html#axzz1phkdKDmo.

scope from policymakers, which is lacking in Vietnam. All laws in the legal framework are linked and the government considers the framework to be perfectly constructed. In fact, there exist conflicts among laws and the government needs to work hard at reducing these struggles within the legal system.

Although efforts are being made to improve the legal framework on agricultural land, the definition of land ownership remains unclear, resulting in problems in establishing land rights.

Since 2007, when Vietnam became a member of the WTO, the government has adjusted and amended the legal framework not only to meet the demands of economic development, but also to meet the requirements of WTO membership.

The second challenge is the reform of civil service. Farmers lack knowledge about laws and their rights; therefore, civil servants need to explain their rights and obligations according to current legal documents. However, the performance and quality of civil servants are not good enough. This has led to the wrong or inadequate explanation of government policy, farmers' rights, and their obligations. By using knowledge about the law, many government officials exploit the inadequate knowledge of farmers on public policies, asking them to pay extra money in order to process farmers' applications for borrowing money from the bank speedily. Recognising the current bad situation of civil service, the Public Administration Reform action plan of the government put the improvement of civil service cadres as one of the four components of the program.

To improve the quality of civil servants, competency-based training curricula have been introduced and implemented by the Ministry of Home Affairs. However, the ministry only focused on several special types of positions and work, leading purely to competence-based training in special cases being developed. This has led to the inefficient government spending on Public Administration Reform. In fact, the government has borrowed USD 15 million from the Asian Development Bank for retraining government officials. However, this programme was implemented slowly due to the lack of competency requirements for the design of competence-based training.

The third challenge is the institutional development for decentralisation. As people become wealthier, the rise of household incomes provides opportunities for increasing choice in households' consumption and investment. Parallel with the improving of household living standards, the party and government have committed themselves to developing institutions that allow for greater participation of people in the control over collective resource allocation, through the development of so-called 'grassroots' democracy. Decentralisation has become a popular strategy for reorganisation of government. Proponents of decentralisation assume that delegating the power to local units of government can promote superior efficiency in government, as local units have a better knowledge of local needs and can meet these needs with lower costs and have them be more responsive. Under certain conditions, decentralisation can bring such benefits; however, it is also true that decentralisation cannot achieve responsiveness to local community needs without the participation of the people. Experience in Vietnam shows that it is easy to decentralise power, but it is very difficult to decentralise power in a way that promotes cost-efficiency and effectiveness of services. Hence, managing power in a decentralised system is both difficult and important.

To ensure that the needs and concerns of the broad mass of the population are reflected in government programmes and that officials are accountable for performance, there is a need for the continuing development of the role of representative political institutions at grassroots level. In addition, the potential importance of incentives and sanctions, resulting from public monitoring of performance, cannot be overemphasised. In education and health especially, the problem of incentives is of critical importance. In many areas of services provisions, there are fundamental conflicts of interest that arise from service providers' desire to enhance incomes and public needs.

This is a challenging task. Even in societies with a long tradition of decentralised local government, active participation of the population is often limited, with control over local decision-making falling into the hands of local interest groups and political cliques. It is therefore not surprising that progress in making 'grassroots' democracy operational has been slow. The other reason for the slow progress is the low awareness of new rights that afford the public the opportunity to monitor service delivery. Surveys have indicated the fact that not many people at the grassroots level understand the Ordinance on Grassroots Democracy, issued in 2007, in which their rights are stipulated in the ordinance.[56] Two of the three key rights of local people (being informed, involved in discussions, and ownership of projects) are usually ignored. The right to be involved in discussions has been mentioned by many local leaders, but in practice, local people have rarely been given appropriate opportunities to participate in discussions and express their opinions to influence choices

---

[56] As shown from interviews with the local people in the Dinh An commune, Tra Cu district, Tra Vinh province.

in local development. Community participation has been limited, in most cases, to offering comments on the list of public investment projects to be financed by communes already identified by the local leaders. However, positive results of decentralisation can be observed in particular in more progressive provinces, which have been able to experiment with reform; it is hoped that such positive experiences, where they have proved successful, will be emulated in follower provinces. Also, even where local participation in decision making has not developed, decentralisation has allowed the mix of programmes and policies to be adjusted to local conditions.

The final challenge is the reform of public finance management. With support from international donors such as the World Bank and the United Nations Development Programme, the government has put significant efforts into the reform of public management; however, the outcomes do not properly match the desires of the donors. At the local level, predictability of funds for commitment to budgeted expenditures is very limited, because the estimation of some revenue sources is often inaccurate, while other sources are not predictable. This situation impacts negatively on budgeting and execution and creates difficulties for managing the budget.

Budgeting at both national and local levels have not been sufficiently based on information regarding the performance of state management agencies and the impacts/results of public investment programmes/projects and policies. Reliable, timely and comprehensive data on the impacts/results of many public investment programmes/projects and policies are unavailable. Staff in some districts and provinces use statistics offices as the key channel for data (DANIDA, 2009). However, the data from statistics offices is not sufficient for making analysis of the outcomes and impacts of Public Investment Programmes. It is often not clear why a project is being financed by government, when it could be readily undertaken by the private sector. Lack of clear and systematic criteria for selection of public investment projects/programmes and poor coordination in planning have contributed to the low efficiency of public investment. For the twelve adjacent coastal provinces, there are nine airports (eight in operation and one under construction). Many cement plants and cane sugar mills have been built in a given province or in adjacent provinces, causing shortages of input materials and excess capacity. Wrong selection of projects/programmes to be financed by public funding may also result from bribes from those entities that will benefit from these projects/programmes. This type of corruption will not be exposed even with careful auditing.

In fact, decentralisation of decision-making is not without problems, which include as follows:

(i) Weaknesses in the coordination of provincial level investment have sometimes led to excess investment in 'popular' activities and a chaotic pattern of development.
(ii) There may be dangers of polarisation as more successful provinces are able to pull further ahead of the pack.
(iii) An additional problem that has now widely recognised in relation to the movement toward decentralised control of essential social services is that while decentralisation can promote greater responsiveness to local needs and can enhance efficiency of operations, it can also produce unaccountable and corrupt local institutions that respond inadequately to public needs.

Priorities of public policy on further social capital development

First, working on the Land Law is a high priority. The proposal of the amendment of Land Law 2005 should be published on the website of the government and the proposal also needs to be published in local newspapers to collect comments and feedback from citizens at the local level before submission to the National Assembly. During the revision of Land Law, related laws such as Law on Construction, Law on Investment, and Law on Real Estate business should be revised to ensure consistency.

Moreover, improvement of contractual and non-contractual obligations should be addressed systematically by the government. The enforcement of contractual obligations is very important to agricultural development. This is an area in which there is much room for improvement. The existence of many laws/ordinances regulating contractual matters, including Civil Code and the Ordinance on Economic Contracts and the Commercial Law, has led to different interpretations in applying laws, which increases business risks. This has also led to a situation where whole buyers impose lower prices in signing contracts with farmers who do not have much knowledge about the complicated system of laws related to buying farmers' products. Consequently, benefits gained by whole buyers are much higher than benefits gained by farmers.

Practically, there are still two separate laws regulating civil contracts, the Civil Code (amended 2005) and the Ordinance on Economic Contracts. Along with these two key legal documents, contracts in the agricultural sector are regulated by other laws that include the Commercial Law, the Law on Credit institutions, and the Law on Insurance. Ordinance of Economic Contracts should be removed; the Civil Code and Commercial Law should be amended in order to improve the provisions on economic contracts. Further efforts should be put on the improvement of the legal framework for non-contractual obligations, such as compensation to the affected for the damage of the assets belonging to others.

Second, civil service development is essential for further agricultural development during the next decade. Vietnam is now an official member of the WTO, so agricultural development should be put along with the development of not only domestic demand but also further integration into the world economy. Therefore, civil servants need to clearly understand either government documents or WTO policies in order to effectively guide their customers. The quality of civil servants needs to be improved from that of the past as a result of these new circumstances.

To motivate civil servants and ensure the best people are in the right posts, recruitment and promotion should be merit-based. The new Law on Public Officials and Civil Servants has integrated many positive ideas for civil service reform. However, to ensure the law is effective, under the law, legal documents required for its implementation should be enacted incorporating the merit-based principle. Job descriptions and performance appraisal criteria should be developed in all government agencies.

Competency requirements must be developed for all types of positions and work, so that the training given to the cadres and civil servants can be competency-based and recruitment and promotion can be merit-based. Gender equality should be taken into account in developing criteria for recruitment and promotion.

The performance of government officials and civil servants should be evaluated based on performance and output indicators, and competencies required to do the job. Recently, a new department (staff performance appraisal) has been established in all Vietnamese ministries/agencies.

Further efforts should be made by all government agencies to establish job descriptions and performance appraisal criteria.

Third, institutional development for decentralised accountability should be done stably and with strong commitment from the government so that the process will be suitably improved. A case can be made that Vietnam has done precisely the correct thing by rapidly decentralising authority over many facets of service provision. However, Vietnam's decentralisation may have proceeded too quickly, ahead of adequately developed regulatory and popular control and monitoring institutions. Developing technically sound and politically workable regulatory institutions is a difficult challenge facing Vietnam during the next decade. The problem is that there is built-in resistance to regulation, as regulation involves getting people to behave in ways they would otherwise prefer not to.

The Vietnamese Communist Party and Government should consider in depth how the party and government bodies should be better designed to ensure that Vietnam's institutions are responsive to the democratic aspirations of the people. Participatory and representative institutions, such as grassroots democracy activities and the election and operation of political bodies (e.g., People's Councils, People's Committees, and the National Assembly) should be designed to ensure greater accountability. Additionally, to improve representative oversights, Vietnam may also need a different kind of regulatory body that, ideally, functions autonomously from local and national units of government.

The final priority is the improvement of public investment efficiency. Despite the fact that the government amended the Law on State Budget 2002, which increased the autonomy of the provincial authorities, decentralisation has not been supported by adequate improvement in local authorities' management capabilities and inter-provincial and regional coordination. In order to improve the quality of public investment projects/programmes, monitoring and evaluation of the projects/programmes should be treated as a high priority task. Baseline data has been provided prior to the start of project activities to provide a basis for evaluating the impact of the project when completed. Additionally, funds from the state budget are allocated to project owners, requiring data on the impact of their projects, which uses funds from the state budget in previous years.

8.4. Conclusion

Agricultural development policies are very important to the d economic development process in Vietnam. The centre of agricultural development in Vietnam is farmers, but farmers are vulnerable under the impact of natural, market and socio-economic shocks. Therefore, the Vietnamese government should build up systematic policies that need to incorporate with other policies related to human and social capital development for either farmers or the rural labour force. Education and health human capitals are the two major elements of human capital, while institutions, social relationships, and inter-family relationships in rural areas are the most significant elements of social capital. Further agricultural development programs in Vietnam should treat human and social capital as focal points. Agricultural development in the current circumstances requires comprehensive policies covering not only domestic demand but also international trade.

Education policy is an important push for agricultural development. Agricultural household members should be retrained in modern skills and knowledge to improve their ability to adapt to new conditions of economic development. Training farmers has two goals: (i) equip farmers with fresh knowledge; (ii) provide young farmers an opportunity to leave agriculture. We cannot expect that older farmers will leave agriculture following training, because they do not have the necessary dynamic in place to leave agriculture; however, young farmers are more self-motivated to leave their farms if they have adequate skills for working in industrial sectors. Moreover, obtaining fresh knowledge will help farmers in applying new technology and modern machines to their work, which may lead to a considerable rise in agricultural productivity.

Besides education policy, human health capital development policy can be seen as a crucial means for furthering agricultural development. Providing better healthcare with financial support from the government needs to be properly implemented. Farmers are among the poorest of the Vietnamese population. They often do not have adequate money to cover healthcare fees and medical expenditures. Therefore, financial support from the government has an important role in increasing human health capital for farmers. This policy, however, requires a large amount of spending from a government under budget constraints. Spending more on healthcare for farmers will constrain government expenditures on other investments; there is therefore a need for estimating effective spending for healthcare from the state budget. Efficient spending on health care infrastructure, such as hospitals and local health care centres, efficient spending on training health professionals and health insurance spending for farmers can be seen as major tasks for the government in meeting the goals of the New Rural Development programme.

Finally, institutional transformation is needed for further agricultural development in Vietnam during the next decade. A sophisticated and comprehensive legal system should be drawn up in order to meet the new circumstances. The system should meet either the needs of further economic transformation or the desires of farmers to improve their living conditions. Therefore, it is required of the government to enhance the capacity of government officials in drafting policies and implementing these policies. Policies based on the needs of farmers require a process that provides opportunities to farmers to participate in public policy formulation, where they can express their desires and rights. Farmers should be seen as a focal point of policies and be given the right to monitor and evaluate the outcomes of policy implementation. The relationship between the government and farmers will in this way be strengthened to achieve the goals of policies. Accordingly, the government can adjust its policies to meet the needs of farmers.

Chapter 9

CONCLUSION

The standard policy prescription for enhancing the productivity of agriculture in a developing country like Vietnam is to first encourage the investment of farmers in their human and social capital, then change the governmental institutions to enable farmers' investment. This thesis therefore analyses the investment of farmers in their health, education, and social relationships in the context of Vietnam's recent agrarian transition. Using the tools of regression analysis, the author has attempted to measure the rate of return of investment in health, education and the social relationships of farmers as it affects their income. Additionally, to measure the effect of local government policy on the performance of agricultural firms at the provincial level, the thesis applied current techniques to estimate the relationship between output of agricultural firms and performance of local government. As in other low-developed countries, the rate of return of investment in education was quite extreme and the rate of investment in health was small. The effect of investment in the social capital of farmers on their income was quite complex due to the complicated nature of the term 'social capital'. Moreover, the effect of the quality of local institutions on performance of agricultural firms at the provincial level is not statistically significant. The relationship was found to be similar to other worldwide research findings yielded when researchers attempted to measure the effect of the quality of institutions on the performance of firms.

The way in which investment activities was assigned to agricultural households in the first stage of transition—in particular, the rights of farmers to invest privately in their land—was clearly crucial both for improving the living standards of farmers and the performance of economic development. However, the heavy reliance on physical investment has raised concerns about horizontal investment, which leads to an unstable increase in agricultural productivity. Vertical investment with a focal point on technology, skills, and the abilities of farmers, as well as quality health farmers are needed for the modern development of agriculture in developing countries. Evidently, in Vietnam, horizontal investment has helped to develop agriculture in the past, while currently, this form of investment is making room of the further development of agriculture. This thesis has first tried to establish if such concerns are borne out by the evidence on how investment in health, education, and social capital affect farmers' income under the new rural development programme introduced by Vietnam in 2000. This was arguably the most crucial task in the country's transformation to market-based agriculture after abandoning the inefficient collective system.

Individual households had to be assigned the rights to invest in their own land. The author used a model of household income to analyse the effect of investment in health and education on their rights of investing in their own land, as expressed by their income. The results are quite similar to the picture many commentators have painted thus far.

Rural development programmes have run through stages since Unification day in 1975, and agriculture was a starting point in the reformation process in 1986. However, agricultural households still rate among the poor income quintile in Vietnam. The situation implies that the government might neglect rural areas when implementing new strategies for economic development. A large amount of Vietnam's labour force is in agriculture and a large percentage of the population live in rural areas. They should therefore be at the centre of the economic development programme rather than on the side lines. The decentralised reform will lead to more investment of farmers in increasing their human and social capital, which will result in a high rate of return on their investment in health, education, and social relationships. The empirical results in this thesis suggest that poor income agricultural households do not have adequate money to simultaneously invest in health and education, and have therefore faced a trade-off between health and education. If farmers invest more in their health they will invest less in their education and their children's schooling. When selecting an option, they will choose to invest more in their children's schooling, with the expected rate of return of investment in health being small.

Combined with readings concerning the history of Vietnam around this time, it is suggested that there are two main reasons for the current situation of rural development in the country. *Firstly*, the formation of pro-reform partnership between farmers and central reformers is a primary reason. The latter political leaders who supported reform were fully aware that the decentralised administration of the government was an obligation of successful transformation of the economy. The desire for transformation was not in fact a governmental concern only, but also reflected the need for replacing the inefficiency of collective agriculture among those who were losing most, i.e., farmers, by creating a more efficient system. Farmer resistance had initiated reforms; however, central government nonetheless played a crucial role in these reforms. Recognising that local government might resist these reforms, the central government encouraged the development of farmers' organisations in order to help shift the balance of local power and used the press to channel complaints and highlight corruption.

Secondly, initial conditions at the time of reform appear to have been favourable for achieving equitable assignment of the ownership of physical capital, so as to encourage the investment of farmers on their physical capital (i.e., human capital). Vietnam's low inequality due to socialist policies in the initial distribution of education can be seen as an ignition for further agricultural development. However, the differential access of farmers to a health infrastructure, such as hospitals in cities and provinces and local health care centres at the time of reform has constrained the development of farmers' human capital.

Additionally, the market for education and health has been developed unequally between cities and local areas and urban and rural areas. Farmers who live near big cities or provincial centres have a better chance of improving their human capital than farmers who live far from such favourable areas. This is a failure of the government in providing equal opportunities for citizens to enhance their human capital. Empirically, better education and health will lead to more productive farmers; in turn, they will be able to exploit their physical capital better. Consequently, the production of agriculture will be increased. Moreover, under the circumstances of deeper integration into the global economy, farmers will need to be retrained with modern knowledge, the likes of which cannot be achieved by farmers who live in remote areas or long distances away from cities or provincial centres. This is another cause of inequality in farmers' incomes in addition to unfavourable conditions of land.

What happened to the investment of farmers in health after the reform?

After the implementation of the reform strategy, the agriculture of Vietnam experienced a significant transformation from a pure socialist to a semi-socialist system. In the new system, Vietnamese agriculture has achieved higher productivity and the living standard of agricultural households has improved considerably. Farmers not only meet their food demand but also have surplus products to sell in the market. This achievement is remarkable, because under the old mechanism, agricultural households did not have enough agricultural foodstuffs such as rice and meat; as such, the Vietnamese government had to import rice from India and China.

Under the new mechanism, agricultural households have more options for investing in their health capital. The increase in farmers' incomes and the availability of health care centres are the two main factors for health investment activity in these households. Increasing household incomes allows farmers to invest more in their health. Based on the household living standard survey, the monthly consumption expenditure per capita for healthcare has increased nearly twofold from 2002 to 2006. Moreover, the government's policy on the development of healthcare centres at different levels also helps farmers in rural areas to meet their health capital investment demands. However, the problem the government faces is that farmers are underestimating the role of their health in the improvement of their productivity.

Therefore, this thesis tried to estimate the effect of farmers' health investment on their productivity, proxied by their income. The empirical results show the low effect of investment in health on income at the individual level by using micro data and are presented in chapter 5. The elasticity between investment in health and income at the individual level is only 2%, much lower than bank interest. The estimated outcome indicates that the economic conditions of individuals are constraining the investment in health of farmers. Although farmers' incomes are increasing, farmers do not invest much in their health. They also have to spend their income on other demands such as tuition fees for their children, maintenance of their house, foodstuffs, electricity, and inputs for their harvest.

Additionally, the thesis aimed to estimate the effect of health investment on famers' learning-by-doing abilities, based on the guidance of endogenous growth theory. Because the author did not have any information or variables about learning-by-doing, the author estimated the effect of past knowledge on current learning-by-doing abilities. The empirical results support the idea that farmers rely greatly on their past knowledge in doing their jobs—just over 80% under the current investment in health strategy. This means that Vietnam's government will struggle in applying new agricultural technology to yield higher productivity in agriculture.

What happened to investment in education of farmers?

Using Hausman and Taylor's instrumental variable regression method (which includes time-invariant and time-variant variables) to an income-education model, the author estimated the rate of return of investment in the education of farmers in the past. The HT/IV method was employed because the OLS and GLS regression methods usually yield a low rate of return and indicate an over-investment in education. This result does not reflect the true relationship between investment in education and income in the case of Vietnam. The empirical result should show an under-investment in the education of farmers. Moreover, the HT/IV estimates the effect of the individual unobservable effect, which might point out the effect of the learning-by-doing abilities of farmers on their income. The empirical results of the HT/IV method demonstrated the situation that farmers in the past were under-investing and the policy implication for the current government is that it should provide more chances for farmers to invest in their learning-by-doing abilities, such as, for example, by making available vocational training courses.

The Mincerian wage equation was transformed to an income-education model, which yielded new empirical results for Vietnamese policymakers in considering new schemes of vocational training for farmers in the future. There is no doubt that vocational training provides farmers the opportunity to obtain new technological knowledge, which is provided by government bodies at different levels. Knowledge in the past played a key role for farmers in their jobs; however, it is not enough for further development in the agricultural sector of Vietnam. The outdated knowledge of farmers is now a limitation in agricultural development in Vietnam's context. Therefore, farmers need to acquire more knowledge. However, empirical results on this topic are limited in the Vietnamese context. Policymakers should also be aware that without further education for farmers, the agricultural sector cannot have the same rate of growth as in the past.

What happened with investment in the social capital of farmers?

Investment in the social capital of farmers is another aspect of investment in the agriculture of Vietnam at the micro level. 'Social capital' is not a new term in the Vietnam research community or the international academic society. However, investment in social capital within the context of economic development in a developing country like Vietnam is a relatively new initiative. Social capital is a multi-faceted economic term; therefore, investment in social capital should offer a spectrum of investment activities to investors. When we wish to measure the rate of return of an investment activity, we usually look at variables that can be measured by money or an index, such as years of schooling. For social capital, we do not have a common index; therefore, this analysis chose to measure investment in social capital using money. Based on the guidance in the literature on economics and sociology, the author classified investment in social capital into two levels: (i) household level and (ii) provincial level.

For household level, expenditures to maintain the inter-family and outer-family relationships were selected as proxies for investment in social capital at the micro level. These variables were not perfect proxies, because we did not have a dataset that included information on investment in social capital and income at the micro level in the context of Vietnam. To overcome the problem, the author employed three waves of VHLSS datasets from 2002, 2004, and 2006 to estimate the rate of return of investment in social capital. The empirical results support the idea that investment in social relationships has a positive effect on the increase of farmers' incomes.

For the provincial level, a stochastic frontier production function model was estimated to show the effect of reformation of local government on the outcomes of representative agricultural firms. The empirical results illustrate a complex picture that does not make it easy to clearly explain these effects. Human development policy of local government does not support the increasing of representative firms in the agricultural sector, and the state-owned enterprises bias policy negatively affects the outcomes of firms. However, not all variable proxies for the reformation of local government support the idea that institutional transformation significantly affects the results of agricultural firms.

What should the Vietnamese government do to foster further agricultural development?

The Vietnamese government has had many successful achievements in the past, and now wishes to further develop its agriculture under the new circumstances. This means new systematic policies need to be built up that are based on empirical results. In the past, the old mechanism was based on socialist planning, while in the *Doi moi* strategy, a market economy was applied in order to encourage the investment of farmers in improving their productivity and incomes. At present, Vietnam's economy is integrating into the international economy deeper than ever before; therefore, the Vietnamese government needs to implement new policies to boost agricultural and rural development. New circumstances require farmers to learn more in order to have higher productivity and incomes and the government plays a key role in supporting farmers in obtaining this new knowledge and skills. Assistance from the government will help farmers to adapt to new conditions and to transfer opportunities into realities.

This thesis contributes minor suggestions to policymakers for gaining further development in agriculture based on its empirical results. *Firstly*, the government should focus on the vocational training of farmers.

With updated skills and knowledge, farmers can not only improve their productivity and incomes, but also have better chances for exiting the agricultural industry. When this happens, the accumulation of agricultural land will be boosted, enabling larger farms to be created. These large farms can apply new technology and be operated easier than small farms, thereby yielding more income as suggested in the literature of agricultural economics. *Secondly*, government at different levels should improve its health care centres. This will help farmers to access healthcare services better than before. The government also should enhance the current health insurance programme to allow farmers to gin more benefits. The support of government via a health insurance programme will assist farmers to invest more in their health. Farmers with better health might have a better chance to earn higher incomes. The increasing of farmers' incomes is the major goal of further agricultural development in the new rural development strategy. *Finally*, reformation of local institutions needs additional development. The performance of local institutions should be monitored and evaluated by the central government in order to have better policies in order to successfully implement the new rural development strategy.

What are the contributions of the thesis?

This dissertation aims to contribute to the current economic literature about the role of investment in health, education, and social capital on the productivity of farmers in Vietnam. This type of household was selected for three reasons. *Firstly*, they play a key role in the early stages of the *Doi moi* strategy. *Secondly*, it is widely observed that in Vietnam, agriculture is affected by a large amount of agricultural households using obsolete and inefficient machinery. This indicates that the increasing productivity of agriculture in Vietnam will be constrained. Therefore, if Vietnam wishes to promote the further development of agriculture, the government needs to provide strong support to this sector. *Thirdly,* the Vietnamese government is implementing a new rural development programme, so there is a need for analysing the current situation of agricultural households' capitals. The core element of the new programme is that the government wishes its farmers to independently decide on suitable investment activities, based on their current needs and assets. Thus, three research questions have been raised and by answering them, the contributions of this thesis are six-fold.

Firstly, this thesis is the first trial that uses an income-health model using VHLSS datasets to measure the rate of return of investment in health on income at the household level with a focus on Vietnamese farmers. In Vietnam, research using VHLSS datasets is not rare, but there are nonetheless only a few studies on agricultural households' incomes. Most research uses households' expenditures, proxied for households' well-being, as independent variables, while most studies focus on explaining the pattern of households' well-being. The major reason for this is that households' incomes are difficult to estimate. This difficulty stems from the unreliable nature of the source of data. The estimation of agricultural households' incomes is more difficult than non-agricultural households', because the former's major source of income comes from self-supply.

Additionally, research into health usually emphasises health status and health indicators, while few studies use income-health models to analyse the effect of health investment on income at the household level. Therefore, this thesis wishes to fill the gap in related economic literature by estimating the relationship between households' incomes and their investment in health.

Secondly, the thesis measures the rate of return of investment in education on agricultural households' incomes in Vietnam using Hausman and Taylor's method. The method requires panel data, while the VHLSS datasets are not real panel data. Only some households are re-interviewed during the waves of collecting data. Therefore, not many researchers choose this method for measuring the rate of return of investment in education. However, Hausman and Taylor's method can estimate the individual unobservable effect that indicates the ability of farmers for learning from each other. Individual unobservable effect estimation can help to establish learning-by-doing for farmers. In fact, using the Hausman and Taylor method will lead to an increase in education coefficient, but the reason behind the increase is not usually well-explained. The author found that using this method can describe the true fact of investment in education in Vietnam, because the empirical results indicate a situation where under-investment is happening in Vietnam's agriculture sector.

Thirdly, the thesis measures the effect of maintenance of inter-family relationships on agricultural households' income. These relationships are very important in the cultural context of Vietnam. It is a tradition within the development of rural Vietnamese societies, but has decreased since the implementation of the *Doi moi* policy. It is difficult to explain the reason for the decrease, but under the development of rural economics, the relationship between rural family members is nonetheless changing. This change might affect the increase of agricultural households' incomes because of the rising mobility of families. However, it is too soon to say whether the change will be positive in the long term. However, at least this change does not harm the development of rural families. In order to capture the effect of the adaptation on the productivity of agricultural households, proxied by income, this thesis wishes to contribute to the current economic literature by estimating this effect. It was found that the evaporation of family relationship does not constrain the development of households' incomes. The finding is similar to the results in other studies in other countries and communities worldwide.

Fourthly, the thesis is the first trial to measure the effect of time use in the bonding of social relationships and its effect on agricultural households' incomes. The allocation of time is a current debate in economic literature and the effect of the allocation on income does not currently hold much attention at the household level. The reason for this is that research in this context is not easy. Time is unobservable and therefore the allocation of time is difficult to estimate. Thus, estimating the effect of allocation of time on income is risky. The outcomes of such estimations may not be reliable due to the lack of appropriate data. In this study, the author has struggled with this situation as well; to overcome this difficulty, a simple method was applied. Using information on time of working, and with an assumption about resting time, the thesis refers all the rest of day time will be allocated to maintenance of the relationships between people. This can be seen as a good way for estimating the effect of bonding social relationships on agricultural households' incomes, but there is nonetheless the need to be very careful in interpreting the results. However, the empirical findings support the idea that bonding social relationships might help farmers raise their incomes; as such, a policy implication in this instance is that the government should support farmers with a place for organising their meetings. It also explains why so many community halls have been built in Vietnam, and furthermore also supports the idea that a criterion for the new rural development programme in Vietnam should be that all communes need to have their own community halls.

Fifthly, the thesis provides a first-time analysis of the relationship between the performance of local government and the revenue of agricultural firms at provincial levels for the case of Vietnam. Most research on firms employ data at firm level and analyse the effect of trade policy on the revenue of firms, but there are not many studies concerning the effect of the quality of local government on the revenue of firms. This situation is happening due to a lack of data. Fortunately, since 2005, USAID has conducted a survey measuring the quality of local government using the opinions of firms. The institution asks firms in all provinces of Vietnam to evaluate the performance of local government; the outcomes of the survey are published annually. The availability of this dataset allowed the author to construct panel data at the provincial level to measure the effect of the performance of local government on agricultural firms' revenues. However, the empirical results of this thesis still require further development in the future.

Finally, the thesis provides comprehensive suggestions on public policy frameworks to the Vietnamese government for further agricultural-rural development, based on the empirical results yielded by the study. The new rural development programme was implemented in 2010, but the preliminary analysis was done over five years from 2004 to 2009. The analysis, however, was only good enough for making the initial decision that the programme needed to be implemented. The future of programme implementation is therefore not systematic. In order to advise the government on further policy implementation, this thesis builds up a policy framework based on its empirical results. Policy implications cover three main areas of the program: rural health, rural education, and rural community.

What further research should be conducted in the future based on the outcomes of the thesis?

Based on the outcomes of the thesis, three potential research topics should be addressed in future studies. The first should focus on the determinants of human and social capital investment behaviour; the second should focus on investment in human and social capital between generations; the third topic should focus on the link between private and public investment in human and social capital.

Firstly, human and social capital investments have been determined by different factors. Each factor has its own effects on the approach to investment of households and individuals. In the first place, some households might choose to invest in human capital first, while others may invest in social capital before human capital. Investors have limited information about the future outcomes of their investments, but they do know that they have to satisfy current consumption in order to gain higher incomes in the future. However, the point here concerns how people spend their resources to invest in human and social capital; what factors will be considered carefully to make final decisions? Social factors such as investment incentive policies from the government, investment environment, or the accessibility of banking systems and private factors such as the availability of financial resources of investors, and the knowledge of investors about investment strategy will be the determinants of human and social capital investment (G. Becker, 1962; D. Romer, 2006; P. Romer, 1989).

Identifying the determinants of investment behaviour will help households overcome their struggle in investing in human and social capital, while helping government to issue appropriate policies for encouraging investment from households. Investments in human and social capital are unavoidable in the life course of individuals; therefore, people need to be aware of encouraging and discouraging factors in order to have efficient and targeted investment. The way of thinking about investing will be determined by external and internal factors, while investment behaviour in human and social capital will be affected mostly by internal factors. The reason for this is simple—everyone is faced with the same social infrastructure, but each person will employ the infrastructure differently based on their own financial availability and accessibility to a banking system.

Secondly, the aspect of investment in human and social capital between generations could be seen as another future research topic (G. S. Becker, 1962). There may be a close relationship between human and social capital; the question in this instance is can investment in human capital lead to a change in social capital? Investment in human capital can be placed into two major categories: investment in education and investment in health. Investment in education will enhance individuals' knowledge and provide them with wider social and professional networks and increased relationships. When individuals' networks and relationships have been widened, they have to spend time and financial resources to maintain the life of these networks and relationships. For example, farmers have to improve their knowledge related to agriculture in order to contribute to their friends or relatives opportunity to increase their income; and then when income has been increased, farmers are willing to learn more from their friends or relatives and share their successful. There is no clear border or causal relationship between investment in human and social capital. Only when we put the two types of investment into a specific context can we see a clear relationship between investment in human and social capital.

Coleman (1988) believes that investment in human capital will create social capital. The author supports this idea, but would like to add that investment in social capital will also enhance the human capital of people. Therefore, investment in human and social capital can be seen as a two-way relationship. When an individual starts with investment in human capital in order to, for example, gain more social capital, he/she will automatically spend more time and financial resources to maintain social capital. Unfortunately, there is no empirical evidence concerning this relationship, and scholars in sociology, economics, and business try to avoid talking about this particular relationship.

However, in the real life, we cannot avoid spending our time and financial resources on human and social capital. The point here is, if we cannot avoid spending resources on them, what should we do in order to better benefit from them? This means that we need to better understand investment in human and social capital and the causal links between them. The particular context of investment in human capital will lead to an increase of social capital and vice versa. If people understand these links clearly, they will be able to create better plans for their lives.

Finally, the link between public and private investment in human and social capital can be an interesting future research topic. Human and social capitals are not only the property of individuals, but also an asset to society. The accumulation of the human capital of an individual will also be the human capital of a nation, which is similar in the case of social capital. Human and social capital can be divided into a micro and macro level (Coleman, 1988; Sandra Franke, 2005). At micro level, we refer to the human and social capital of a specific person, while at the macro level we refer to the aggregation of human and social capital. At the macro level, human capital includes an aggregation of private human capital and public human capital. Private human capital is an individual's skills, knowledge and health, whereas public human capital can be seen as the capacity of a nation in terms of skills, knowledge, and health stock. Therefore, private investment in human capital is investment in education and health, while public investment in human capital is investment in capacity building and public health. The spending of government on the capacity building of a nation is a major source of growth in the modern world. The question here is to what extent does public investment in human capital support private investment in human capital, and vice versa? The link between the two activities remains unclear and awaits further research.

For investment in social capital, the link between private and public investment should be put on the central of modern economics. In fact, social capital is not a clearly defined term; it has a multifaceted nature in terms of definition. Therefore, we cannot expect that the definition of social capital will cover all aspects of social capital. Instead, we only know about some aspects of social capital. Private investment in social capital concerns the time an individual spends on creating and maintaining his/her network, while public investment in social capital is a part of state budget spending on providing the right environment for the development of communities. Public investment in social capital also includes the time that the government contributes to making and implementing policies related to the development of communities. This means that private and public investment in social capital should go hand in hand. Individuals cannot create and maintain their networks if the government does not provide a legal framework for doing so. Additionally, government spending on social capital will direct the investment of its citizens in improving their networks, thereby facilitating the flow of information. Speedy information flow in modern society creates better opportunities for people to increase their productivity (Lucas Jr. & Moll, 2011). The increase of productivity will lead to economic growth. In short, the links between private and public investment in social capital will be an interesting topic, albeit one that will not be easily explained.

It is said that farmers are investing in future generations via building up their children's human and social capital and that this investment will be valuable in terms of further economic growth in Vietnam. This investment in human and social capital will boost the quality of the labour force in the future and will help to promote industrialisation in Vietnam. Vietnam is a developing country; therefore, Vietnam should take the opportunity to learn from the development of other more advanced countries. Agricultural development in Vietnam will provide a foundation for the development of other industries, which can be seen as a basis for successful industrialisation. Additionally, Vietnam is now becoming a middle-income country; in order to move forward, it needs to develop more agricultural activity through the application of advanced technology, which in turn requires more skilled labourers. Skilled labourers, in turn, are created through education and training.

REFERENCES

Acemoglu, D. (2009). *Introduction to modern economic growth*: Princeton University Press.

ADB. (2010). *Key indicators for Asia and Pacific 2010* (Vol. 41). Mandaluyong City, Philippines: Asian Development Bank.

ADB. (2010). *Microfinance Assessment of ADB TA-7499-VIE: Developing Microfinance Sector in Vietnam.* Hanoi, Vietnam: Asian Development Bank.

Ahlroth, S., Björklund, A., & Forslund, A. (2005). The output of the Swedish education sector. *Review of Income and Wealth, 43*(1), 89–104.

Akram-Lodhi, A. H. (2004). Are 'Landlords Taking Back the Land': An Essay on the Agrarian Transition in Vietnam. *European Journal of Development Research, 16*(4), 757–789.

Amemiya, T., & MaCurdy, T. E. (1986). Instrumental-variable estimation of an error-components model. *Econometrica*(54), 869–880.

Anderson, J. A., & Feder, G. (2007). Agricultural Extension. In R. Evenson & R. Pingali (Eds.), *Handbook of agricultural economics* (Vol. 3, pp. 2344–2381): Elsvier.

Arrow, K. J. (1999). Observations on social capital. In P. Dasgupta & I. Serageldin (Eds.), *Social Capital: A Multfaceted Perspective*. Washington D.C: Worldbank.

Athukorala, P. (2009). Economic transition and export performance in Vietnam. *ASEAN Economic Bulletin, 26*(1), 96–114.

Athukorala, P. C., Pham, H. L., & Vo, T. T. (2009). Vietnam. In K. Anderson & W. Martin (Eds.), *Distortions to agricultural incentives in Asia*. Washington DC: World Bank.

Azariadis, C., & Drazen, A. (1990). Threshold externalities in economic development. *Quarterly Journal of Economics, 105*(2), 501–526.

Baldacci, E., Clements, B., Gupta, S., & Cui, Q. (2008). Social Spending, Human Capital, and Growth in Developing Countries. [doi: DOI: 10.1016/j.worlddev.2007.08.003]. *World Development, 36*(8), 1317–1341.

Barro, R. (1991). Economic growth in a cross section of countries. *Quarterly Journal of Economics*, 407–443.

Barro, R. (1997). *Determinants of economic growth: a cross-country empirical study.* Cambridge, M.A: MIT Press.

Barro, R. (1999). Human capital and growth in cross-country regressions. *Swedish Economic Policy Review, 6*(2), 237–277.

Barro, R., & Lee, J. (1993). International comparisons of educational attainment. *Journal of Monetary Economics, 32*(3), 363–394.

Barro, R., & Lee, J. (1996). International measures of schooling years and schooling quality. *American Economic Review, 86*(2), 218–223.

Barro, R., & Lee, J. (2001). International data on educational attainment: updates and implications. *Oxford Economic Papers, 53*(3), 541.

Barro, R. J., & Sala-i-Martin, X. (1995). *Economic Growth.* New York: McGraw-Hill.

Barro, R. J., Sala-i-Martin, X., Blanchard, O. J., & Hall, R. E. (1991). Convergence Across States and Regions. *Brookings Papers on Economic Activity, 1991*(1), 107–182.

Becker, G. (1962). Investment in human capital: a theoretical analysis. *Journal of Political Economy, 70*(S5).

Becker, G. S. (1962). Investment in human capital: a theoretical analysis. *Journal of political economy, 70*(5), 9–49.

Becker, G. S. (1964). *Human Capital: A Theoretical and Empirical Analysis, with Special Reference to Education*. Chicago, USA: The University of Chicago Press.

Becker, G. S., & Chiswick, B. R. (1966). Education and the Distribution of Earnings. *American Economic Review, 56*(1/2), 358–369.

Behrman, J. R., & Deolalikar, A. B. (1988). Health and nutrition. *Handbook of development economics, 1*, 631–711.

Benhabib, J., & Spiegel, M. (1994). The role of human capital in economic development evidence from aggregate cross-country data. *Journal of Monetary Economics, 34*(2), 143–173.

Ben-Porath, Y. (1967). The production of human capital and the life cycle of earnings. *Journal of political economy, 75*(4), 352–365.

Ben-Porath, Y. (1967). The production of human capital and the life cycle of earnings. *Journal of political economy, 75*(4), 352–365.

Beresford, M. (1985). Household and collective in Vietnamese agriculture. *Journal of Contemporary Asia, 15*(1), 5–36.

Beresford, M. (1990). Vietnam: socialist agriculture in transition. *Journal of Contemporary Asia, 20*(4), 466–486.

Beresford, M. (1990). Vietnam: socialist agriculture in transition. *Journal of Contemporary Asia, 20*(4), 466–486.

Beresford, M. (2008). Doi Moi in review: The challenges of building market socialism in Vietnam. *Journal of Contemporary Asia, 38*(2), 221–243.

Beugelsdijk, S., & Van Schaik, T. (2005). Social Capital and growth in European Regions: an empirical Test. *European Journal of Political Economy, 21*(2), 301–324.

Binswanger, H. (1986). Agricultural mechanization: a comparative historical perspective. *World Bank Research Observer, 1*(1), 27.

Binswanger, H., & Pingali, P. (1988). Technological priorities for farming in sub-Saharan Africa. *World Bank Research Observer, 3*(1), 81.

Bourdieu, P. (1986). Forms of Capital. In J. G. Richardson (Ed.), *Handbook of Theory and Research for the Sociology of Education* (pp. 241–260). Westport, CT: Greenwood Press.

Bowles, S., & Gintis, H. (1998). Is equality passé? Homo reciprocans and the future of egalitarian politics. *Boston Review, 23*(6), 1–27.

Bowman, M. (1969). Economics of Education. *Review of Educational Research, 39*(5), 641.

Breusch, T. S., Mizon, G. E., & Schmidt, P. (1989). Efficient estimation using panel data. *Econometrica*, 695–700.

Brischetto, A., & Voss, G. (1999). A structural vector autoregression model of monetary policy in Australia. *Reserve Bank of Australia Research Discussion Paper, 11*.

Bulow Jeremy, K. R. (1989). Sovereign Debt: Is to Forgive to Forget? [Academic]. *American Economic Review, 79*(1), 8.

Callison, C. S. (1983). *Land-to-the-tiller in the Mekong Delta: economic, social, and political effects of land reform in four villages of South Vietnam* (Vol. 23): Center for South and Southeast Asia Studies, University of California (Berkeley, CA and Lanham, MD).

Campbell, B. (1974). Social change and class formation in a French West African State. *Canadian Journal of African Studies*, 285–306.

Chakraborty, B., & Gupta, M. R. (2009). Human capital, inequality, endogenous growth and educational subsidy: A theoretical analysis. [doi: DOI: 10.1016/j.rie.2009.03.001]. *Research in Economics, 63*(2), 77–90.

Che, N., Kompas, T., & Vousden, N. (1999). Market reform and Vietnamese agriculture. *Vietnam and the East Asian Crisis, Edward Elgar, UK*.

Che, T. N., Kompas, T., & Vousden, N. (2006). Market reform, incentives and economic development in Vietnamese rice production. *Comparative economic studies, 48*(2), 277–301.

Chi, W. (2008). The role of human capital in China's economic development: Review and new evidence. [doi: DOI: 10.1016/j.chieco.2007.12.001]. *China Economic Review, 19*(3), 421–436.

Chou, S. Y., Liu, J. T., Grossman, M., & Joyce, T. (2003). Parental Education and Child Health: Evidence from a Natural Experiment in Taiwan. *American Economic Journal: Applied Economics, 2*, 33–61.

Coleman, J. (1990). *Foundations of Social Theory.* Cambridge: Harvard University Press.

Coleman, J. S. (1988). Social capital in the creation of human capital. *American Journal of Sociology, 94*, s95–s120.

Coleman, J. S. (2007). Social capital in the creation of human capital (Vol. 94, pp. S95): UChicago Press.

Conning, J., & Udry, C. (2007). Rural financial markets in developing countries. In R. Evenson & P. Pingali (Eds.), *Handbook of agricultural economics* (Vol. 3): Elsevier.

Côté, S., & Healy, T. (2001). *The well-being of nations: The role of human and social capital*: OECD.

Dagum, C., & Slottje, D. (2000). A new method to estimate the level and distribution of household human capital with application. *Structural Change and Economic Dynamics, 11*(1–2), 67–94.

DANIDA. (2009). *Public Finance Management Assessment, Planning/ M&E review in five provinces: Dien Bien, Lao Cai, Lai Chau, Dak Lak and Dak Nong.* Hanoi.

Dasgupta, P. (1999). Economic Progress and the Idea of social capital. In P. Dasgupta & I. Serageldin (Eds.), *Social capital: A multifaceted perspective.* Washington, D.C: World bank.

Davies, A. (2001). *But we knew that already!: A study into the relationship between social capital and volunteering.* Paper presented at the Home Start, Sheffield.

Di Maria, C., & Stryszowski, P. (2009). Migration, human capital accumulation and economic development. [doi: DOI: 10.1016/j.jdeveco.2008.06.008]. *Journal of Development Economics, 90*(2), 306–313.

Di Pietro, G., & Pedace, L. (2008). Changes in the returns to education in Argentina. *Journal of Applied Economics, 11*, 259–279.

Diaz-Bonilla, E., & Robinson, S. (2010). Macroeconomics, Macrosectoral Policies, and Agriculture in Developing Countries. In P. L. Pingali & R. E. Evenson (Eds.), *Handbook of agricultural economics* (pp. 3035–3213): Elsevier.

Dinda, S. (2008). Social capital in the creation of human capital and economic growth: A productive consumption approach. [doi: DOI: 10.1016/j.socec.2007.06.014]. *Journal of Socio-Economics, 37*(5), 2020–2033.

Dixon, C. (2003). Developmental lessons of the Vietnamese transitional economy. *Progress in Development Studies, 3*(4), 287–306.

Dolan, P. (2000). The measurement of health-related quality of life for use in resource allocation decisions in health care. *Handbook of health economics, 1,* 1723–1760.

Dow, G., Olewiler, N., & Reed, C. (2005). The transition to agriculture: Climate reversals, population density, and technical change. *Population Density, and Technical Change (January 2005)*.

Dublin, L., & Lotka, A. (1930). The money value of a man. *American Journal of Nursing, 30*(9), 1210.

Eisner, R. (1985). The total incomes system of accounts. *Survey of Current Business, 65*(1), 24–48.

Eisner, R. (1988). Extended accounts for national income and product. *Journal of Economic Literature, 26*(4), 1611–1684.

Eisner, R. (1989). *The total incomes system of accounts*: University of Chicago Press.

Engel, E. (1883). *Der werth des menschen*: L. Simion. Cited in Le *et. al* (2005).

Estudillo, J. P., & Otsuka, K. (2010). Rural Poverty and Income Dynamics in Southeast Asia. In P. L. Pingali & R. E. Evensson (Eds.), *Handbook of Agricultural Economcis* (Vol. 4). Amsterdam, Netherland: Elsevier.

Farr, W. (1852). Equitable taxation of property. *Journal of Royal Statistics, 16*, 1–45.

Fenwick, A., & Figenschou, B. (1972). The effect of Schistosoma mansoni infection on the productivity of cane cutters on a sugar estate in Tanzania. *Bulletin of the World Health Organization, 47*(5), 567.

Fforde, A. (2008). Vietnam's informal farmers' groups: Narratives and policy implications. *Suedostasien Aktuell—Journal of Current Southeast Asian Affairs, 1*, 3–36.

Fleisher, B., Li, H., & Zhao, M. Q. Human capital, economic growth, and regional inequality in China. [doi: DOI: 10.1016/j.jdeveco.2009.01.010]. *Journal of Development Economics, In Press, Corrected Proof.*

Fogel, R. (1991). The conquest of high mortality and hunger in Europe and America: Timing and mechanisms. In P. Higonnet, D. S. Landes & H. Rosovsky (Eds.), *Favorites of fortune: technology, growth, and economic development since the Industrial Revolution* (pp. 35–71). Cambridge, MA: Harvard University Press.

Fogel, R. W. (1994). *Economic growth, population theory, and physiology: The bearing of long-term processes on the making of economic policy*: National Bureau of Economic Research.

Fogel, R. W., & Wimmer, L. T. (1992). Early indicators of later work levels, disease, and death: National Bureau of Economic Research Cambridge, Mass., USA.

Foster, A., & Rosenzweig, M. (2008). Economic development and the decline of agricultural employment. In T. P. Schultz & J. Strauss (Eds.), *Handbook of development economics* (Vol. 4, pp. 3051–3083): Elsevier.

Foville, A. D. (1905). Ce que c'est la richesse d'un peuple. *Bulletin de l'Istitut International de Statistique, 14*(3), 62–74. Cited in Trinh et. al. (2005).

Franke, S. (2005). *Measurement of Social Capital: Reference document for Public Policy Research, Development, and Evaluation* (PDF): Policy Research Initiative, Canada.

Franke, S. (2005). *Measurement of social capital: Reference document for public policy research, development, and evaluation*: Policy Research Initiative.

Fukuyama, F. (1995). Social capital and the global economy. *Foreign Affairs, 74*(5), 89–103. Quoted in Halpern, D. (1999) *Social capital: the new golden goose*. Faculty of Social and Political Sciences, Cambridge University. Unpublished review.

Gardner, R. (1998). Unobservable individual effects in unbalanced panel data. *Economics letters, 58*(1), 39–42.

Gavious, I., & Russ, M. (2009). The valuation implications of human capital in transactions on and outside the exchange. [doi: DOI: 10.1016/j.adiac.2009.09.004]. *Advances in Accounting, 25*(2), 165–173.

Gemmell, N. (2009). Evaluating the impacts of human capital stocks and accumulation on economic growth: some new evidence. *Oxford Bulletin of economics and Statistics, 58*(1), 9–28.

Glaeser, E., Laibson, D., & Sacerdote, B. (2000). The Economic Approach to Social Capital. Retrieved NBER working paper 7728: http://papers.nber.org/papers/W7728

Goldman, D. P., & Smith, J. P. (2002). *Can patient self-management help explain the SES health gradient?* Paper presented at the Proceedings of the National Academy of Sciences.

Graham, J., & Webb, R. (2005). Stocks and depreciation of human capital: New evidence from a present-value perspective. *Review of Income and Wealth, 25*(2), 209–224.

Granato, J., Inglehart, R., & Leblang, D. (1996). The effect of cultural values on economic development: theory, hypotheses, and some empirical tests. *American Journal of Political Science, 40*(3), 607–631.

Greiner, A. (2008). Fiscal policy in an endogenous growth model with human capital and heterogenous agents. [doi: DOI: 10.1016/j.econmod.2007.10.006]. *Economic Modelling, 25*(4), 643–657.

Griliches, Z. (1977). Estimating the returns to schooling: Some econometric problems. *Econometrica: Journal of the Econometric Society*, 1–22.

Griliches, Z. (1997). Education, human capital, and growth: a personal perspective. *Journal of Labor Economics*, 330–344.

Griliches, Z., Hall, B. H., & Hausman, J. A. (1978). Missing data and self-selection in large panels. *Annales de l'INSEE*, 137–176.

Grootaert, C., & Van Bastelaer, T. (2001). Understanding and measuring social capital: a synthesis of findings and recommendations from the social capital initiative. *Social Capital Initiative Working Paper, 24*.

Grossman, M., & Benham, L. (1974). Health, hours and wages. *Economics of health and medical care*, 205–233.

GSO. (2010). *General Statiscal Yearbook*. Hanoi: Statistics Press.

Gyimah-Brempong, K., & Wilson, M. (2004). Health human capital and economic growth in Sub-Saharan African and OECD countries. [doi: DOI: 10.1016/j.qref.2003.07.002]. *Quarterly Review of Economics and Finance, 44*(2), 296–320.

Hans, W., & Nilsson, E. (2005). Measuring Enterprises' Investment in Social Capital: A Pilot Study. *Regional Studies, 39*(8), 1079–1094.

Hanushek, E., & Kimko, D. (2000). Schooling, labor-force quality, and the growth of nations. *American Economic Review, 90*(5), 1184–1208.

Hausman, J. A., & Taylor, W. E. (1981). Panel data and unobservable individual effects. *Econometrica*, 1377–1398.

Heckman, J. J. (2005). China's human capital investment. [doi: DOI: 10.1016/j.chieco.2004.06.012]. *China Economic Review, 16*(1), 50–70.

Helliwell, J. (1996). Economic growth and social capital in Asia. from National Bureau of Economic Research Cambridge, Mass., USA: Available at SSRN: http://ssrn.com/abstract=4696

Helliwell, J., & Putnam, R. (1995). Economic growth and social capital in Italy. *Eastern Economic Journal, 21*(3), 295–307.

Hill, A., & Mamdani, M. (1989). Operational guidelines for measuring health through household surveys. *London: Centre for Population*

Studies (School of Hygiene and Tropical Medicine, University of London).

Huffman, W., & Orazem, P. (2007). Agriculture and human capital in economic growth: Farmers, schooling and nutrition. In R. Evenson & P. Pingali (Eds.), *Handbook of agricultural economics* (Vol. 3, pp. 2281–2341): Elsevier.

Huffman, W. E. (2001). Human Capital: Education and Agriculture. In G. R. B. Gardner (Ed.), *Handbook of Agricultural Economics* (Vol. 1): Elsevier Science.

Huffman, W. E., & Orazem, P. F. (2007). Agriculture and Human Capital in Economic Growth: Farmers, Schooling and Nutrition. In R. Evenson & P. Pingali (Eds.), *Handbook of Agricultural Economics* (Vol. Volume 3, pp. 2281–2341): Elsevier.

Huggett, M., Ventura, G., & Yaron, A. (2006). Human capital and earnings distribution dynamics. [doi: DOI: 10.1016/j.jmoneco.2005.10.013]. *Journal of Monetary Economics, 53*(2), 265–290.

Hung, D. V., & Tam, L. T. (2010). *Rural financial market: Fact diagnostics and the main obstacles of financial supporting policies to informal farmers' voluntary economic groups*. Paper presented at the Policy Dialogue 'Promotion to Farmers' voluntary economic associations in the new rural development.

Iliffe, J. (1979). *A modern history of Tanganyika.* Cambridge, England: Cambridge University Press.

Inglehart, R. (1994). The Impact of Culture on Economic Development: Theory, Hypotheses, and Some Empirical Tests. *Department of Political Science, University of Michigan, Ann Arbor.*

Islam, N. (1995). Growth Empirics: A Panel Data Approach. *Quarterly journal of economics, 110*(4), 1127–1170.

Jacobs, B. (2007). Real options and human capital investment. [doi: DOI: 10.1016/j.labeco.2007.06.008]. *Labour Economics, 14*(6), 913–925.

Jamison, D. T., & Lau, L. J. (1982). *Farmer education and farm efficiency*: Published for the World Bank.

Jarvis, L., & Vera-Toscano, E. (2004). Seasonal adjustment in a market for female agricultural workers. *American Journal of Agricultural Economics, 86*(1), 254.

Jeong, B. (2002). Measurement of human capital input across countries: a method based on the laborer's income. [doi: DOI: 10.1016/S0304-3878(01)00194-8]. *Journal of Development Economics, 67*(2), 333–349.

Johnson, D. (1997). Agriculture and the Wealth of Nations. *American Economic Review, 87*(2), 1–12.

Jorgenson, D., & Fraumeni, B. (1989). The Accumulation of Human and Non-Human Capital, 1948–1984. In R. Lipsey & H. Tice (Eds.), *The Measurement of Savings, Investment and Wealth* (pp. 272–282): The University of Chicago Press, Chicago.

Jorgenson, D., & Fraumeni, B. (1992). The Output of the Education Sector. In Z. Griliches (Ed.), *Output measurement in the service sectors* (pp. 303–338): University Of Chicago Press.

Judson, R. (2002). Measuring human capital like physical capital: what does it tell us? *Bulletin of economic research, 54*(3), 209–231.

Jungsoo, P. (2006). Dispersion of human capital and economic growth. [doi: DOI: 10.1016/j.jmacro.2004.09.004]. *Journal of Macroeconomics, 28*(3), 520–539.

Keane, M. P., & Wolpin, K. I. (1997). The career decisions of young men. *Journal of Political Economy, 105*(3), 473–522.

Kendrick, J., Lethem, Y., & Rowley, J. (1976). *The formation and stocks of total capital*: National Bureau of Economic Research.

Kerkvliet, B. J. T. (2006). Agricultural land in Vietnam: Markets tempered by family, community and socialist practices. *Journal of Agrarian Change, 6*(3), 285–305.

Kerkvliet, B. J. T., & Selden, M. (1998). Agrarian transformations in China and Vietnam. *China Journal*, 37–58.

Kiker, B. (1966). The historical roots of the concept of human capital. *Journal of Political Economy, 74*(5), 481–499.

Knack, S., & Keefer, P. (1997). Does Social Capital Have An Economic Payoff? A Cross-Country Investigation*. *Quarterly journal of economics, 112*(4), 1251–1288.

Koman, R., & Marin, D. (1999). Human capital and macroeconomic growth: Austria and Germany, 1960–1997: an update. Institute for Advanced Studies, University of Munich.

Kompas, T. (2004). Market reform, productivity and efficiency in Vietnamese rice production, *International and Development Economics Working Papers.*

Kompas, T., Che, T. N., Nguyen, H. T. M., & Nguyen, H. Q. (2012). Productivity, Net Returns, and Efficiency: Land and Market Reform in Vietnamese Rice Production. *Land Economics, 88*(3), 478–495.

Krishna, A. (1999). Creating and harnessing social capital. In P. Dasgupta & I. Serageldin (Eds.), *Social Capital: A Multfaceted Perspective*. Washington D.C: Worldbank.

Krueger, A., & Lindahl, M. (2001). Education for growth: why and for whom? *Journal of Economic Literature, 39*(4), 1101–1136.

Kyriacou, G. (1991). Level and growth effects of human capital: a cross-country study of the convergence hypothesis. *Economic Research Reports*, 19–26.

Lam, W. (1998). *Governing irrigation systems in Nepal: institutions, infrastructure, and collective action*: Oakland (California): ICS Press, Institute for Contemporary Studies.

Lam, W. (1998). *Governing irrigation systems in Nepal: institutions, infrastructure, and collective action*: Oakland (California): ICS Press, Institute for Contemporary Studies.

Le, T. D., & Pham, H. T. (2009). Construction of the panel data: Vietnam Household Living Standard Surveys (VHLSS) 2002–2006. Unpublished Working Paper.

Leung, S. (2009). Banking and financial sector reforms in Vietnam. *ASEAN Economic Bulletin, 26*(1), 44–57.

Levine, R., & Renelt, D. (1992). A sensitivity analysis of cross-country growth regressions. *American Economic Review, 82*(4), 942–963.

Lewis, W. A. (1966). *Development planning*: Routledge.

Li, H., Fraumeni, B., Liu, Z., & Wang, X. (2009). Human Capital in China. *NBER Working paper.*

Lin, J. Y. (2009). *Economic development and transition: thought, strategy, and viability*: Cambridge University Press Cambridge.

Lipton, M. (1977). *Why Poor People Stay Poor: Urban Bias in World Development.* Cambridge: Harvard University Press.

Lipton, M. (1993). Urban bias: of consequences, classes and causality. *Journal of Development Studies, 29*(4), 229–258.

Locay, L. (1989). From hunting and gathering to agriculture. *Economic Development and Cultural Change, 37*(4), 737–756.

Long, N. V. (1993). Reform and rural development: Impact on class, sectoral, and regional inequalities. In W. Turley & M. Selden (Eds.), *Reinventing Vietnamese Socialism* (pp. 165–207). Boulder: Westview Press.

López-Bazo, E., & Moreno, R. (2008). Does human capital stimulate investment in physical capital?: Evidence from a cost system framework. [doi: DOI: 10.1016/j.econmod.2008.04.008]. *Economic Modelling, 25*(6), 1295–1305.

Machlup, F. (1984). *The economics of information and human capital*: Princeton university press.

Mahbubani, K. (1995). The Pacific Way. *Foreign Affairs, 74*(1), 100–111.

Mankiw, N., Romer, D., & Weil, D. (1992). A contribution to the empirics of economic growth. *Quarterly Journal of Economics, 107*(2), 407–437.

Mankiw, N. G., Romer, D., & Weil, D. N. (1992). A Contribution to the Empirics of Economic Growth. *Quarterly Journal of Economics, 107*(2), 407–437.

Marceau, N., & Myers, G. (2006). On the Early Holocene: Foraging to Early Agriculture*. *The Economic Journal, 116*(513), 751–772.

Mauro, P. (1995). Corruption and growth. *Quarterly Journal of Economics, 110*, 681–712.

Mendelsohn, R. (2007). Past Climate Change Impacts on Agriculture. In R. Evenson & P. Pingali (Eds.), *Handbook of agricultural economics* (Vol. 3, pp. 3009–3031): Elsevier.

Menon, J. (2009). Managing Success in Vietnam: Macroeconomic Consequences of Large Capital Inflows with Limited Instruments. *ASEAN Economic Bulletin, 26*(1), 77–95.

Mincer, J. (1958). Investment in human capital and personal income distribution. *Journal of Political Economy, 66*(4), 281–302.

Mincer, J. (1974). *Schooling, Experience, and Earnings*. New York: National Bureau of Economic Research.

Mincer, J. (1974). Schooling, Experience, and Earnings *Human Behavior & Social Institutions, 2*.

Moock, P. R., Patrinos, H. A., & Venkataraman, M. (2003). Education and earnings in a transition economy: the case of Vietnam. *Economics of education review, 22*(5), 503–510.

Morand, O. (2002). Evolution through revolutions: Growing populations and changes in modes of production. *University of Connecticut, mimeo*.

Mulligan, C., & Sala-i-Martin, X. (1997). A labor income-based measure of the value of human capital: An application to the states of the United States. *Japan and the World Economy, 9*(2), 159–191.

Mundlak, Y., Larson, D., & Butzer, R. (2002). *Determinants of agricultural growth in Indonesia, the Philippines, and Thailand*: World Bank, Development Research Group, Rural Development.

Nadav, C. (1996). Nutritional thresholds and growth. *Processed, Department of Economics, Ben-Gurion University, Israel.*

Nehru, V., & Swanson, E. (1995). A New Database on Human Capital Stock in Developing and Industrial Countries: Sources, Methodology and Results. *Journal of Development Economics, 46*(2), 379–401.

Newbury, C., & Kanya-Forstner, A. (1969). French Policy and the Origins of the Scramble for West Africa. *Journal of African History, 10*(2), 253–276.

Nguyen, N. N. (2004). Trends in Education sector. In P. Glewwe, N. Agrawal & D. Dollar (Eds.), *Economic Growth, Poverty and Household Welfare in Vietnam.* Washington DC: World Bank.

Nguyen, V. S. (2004). The politics of land: Inequality in land access and local conflicts in the red river delta since decollectivization *Social inequality in Vietnam and the challenges to reform* (pp. 270–296): Institute of Southeast Asian Studies.

Nhu, H. V. (2003). *Situation and factors related to the health care quality in a mountainous community health centre*: Ministry of Health, Hanoi School of Public Health.

North, D. C., & Thomas, R. P. (1977). The first economic revolution. *Economic History Review, 30*(2), 229–241.

Ohno, K. (2009). Avoiding the middle-income trap: renovating industrial policy formulation in Vietnam. *ASEAN Economic Bulletin, 26*(1), 25–43.

Olson, M. (1982). *The rise and decline of nations*: Yale University Press New Haven, CT.

Olsson, O., & Hibbs Jr, D. A. (2005). Biogeography and long-run economic development. *European Economic Review, 49*(4), 909–938.

Ostrom, E. (1990). *Governing the commons: The evolution of institutions for collective action*: Cambridge Univ Pr.

Ostrom, E. (1992). *Crafting Institutions for Self-Governing Irrigation Systems*. San Francisco: ICT Press.

Ostrom, E. (1994). Constituting Social Capital and Collective Action. *Journal of theoretical politics, 6*(4), 527.

Ostrom, E. (1999). Social capital: a fad or a fundamental concept. In P. Dasgupta & I. Serageldin (Eds.), *Social capital: A multifaceted perspective* (pp. 172–214). Washington DC: Worldbank.

Ostrom, E., Gardner, R., & Walker, J. (1994). *Rules, games, and common-pool resources*: University of Michigan Press.

Otsuka, K., Estudillo, J. P., & Sawada, Y. (2009). Toward a new paradigm of farm and nonfarm linkages in economic development. In K.

Otsuka, J. P. Estudillo & Y. Sawada (Eds.), *Rural Poverty and Income Dynamics in Asia and Africa*. London: Routledge.

Packard, L. A. T. (2006). *Gender dimensions of Viet Nam's comprehensive macroeconomic and structural reform policies*: UNRISD.

Passey, A. (2000). *Social Capital: embeddedness and autonomy*. Paper presented at the ISTR.

Paul Schultz, T. (2003). Wage rentals for reproducible human capital: evidence from Ghana and the Ivory Coast. [doi: DOI: 10.1016/j.ehb.2003.08.004]. *Economics & Human Biology, 1*(3), 331–366.

Perri, T. J. (2003). The cost of specialized human capital. [doi: DOI: 10.1016/S0272-7757(02)00093-6]. *Economics of Education Review, 22*(4), 433–438.

Petty, W. (1690). *Political Arithmetick*: Reprinted in Hull (1899).

Pierre, B. (1986). Forms of Capital. In J. G. Richardson (Ed.), *Handbook of Theory and Research for the Sociology of Education* (pp. 241–260). Westport: CT: Greenwood Press.

Pincus, J. (2009). Vietnam: sustaining growth in difficult times. *ASEAN Economic Bulletin, 26*(1), 11–24.

Pitt, M., & Rosenzweig, M. R. (1986). Agricultural prices, food consumption, and the health and productivity of Indonesian farmers. *Agricultural household models: Extensions, applications and policy*, 153–182.

Ponthieux, S. (2004). *The concept of social capital: a critical review.* Paper presented at the 10th ACN Conference, Paris, 21–23 January 2004.

Portes, A. (1998). Social Capital: Its Origins and Applications in Modern Sociology. *Annual Review of Sociology, 24*, 1–24.

Psacharopoulos, G. (1994). Returns to investment in education: A global update. *World development, 22*(9), 1325–1343.

Psacharopoulos, G., & Arriagada, A. (1986). Educational Composition of the Labour Force: An International Comparison, The. *Int'l Lab. Rev., 125*, 561.

Psacharopoulos, G., & Arriagada, A. (1992). The educational composition of the labour force: an international update. *Journal of Educational Planning and Administration, 6*(2), 141–159.

Putnam, R. (1995). Bowling alone: The collapse and Revival of American Community. *Journal of Democracy, 6*(1), 65–78.

Putnam, R. (2000). *Bowling Alone: America's Declining Social Capital.* New York: Simon and Schuster.

Putnam, R. (2007). E Pluribus Unum: Diversity and community in the twenty-first century. *Scandinavian Political Studies, 30*(2), 137.

Putnam, R., Leonardi, R., & Nanetti, R. (1993). *Making Democracy Work: Civic Traditions in Modern Italy.* Princeton, New Jersey: Princeton University Press.

Rama, M., & Võ, V. K. (2009). *Making Difficult Choices: Vietnam in Transition.* Hanoi: The World bank.

Ravallion, M., & Van de Walle, D. (2008). *Land in transition: reform and poverty in rural Vietnam*: World Bank Publications.

Romer, P. (1989). Human capital and growth: theory and evidence. *NBER Working paper, 3173,* 1–49.

Rosen, S. (1977). Human capital: A survey of empirical research. In R. G. Ehrenberg & G. Conn (Eds.), *Research in Labor Economics*: JAI.

Rothschild, M. (1992). Comment on 'output of the education sector'. In Z. Griliches (Ed.), *Output Measurement in the Services Sector* (pp. 339–341). Chicago, USA: The University of Chicago Press.

Routledge, B., & Von Amsberg, J. (2003). Social capital and growth. *Journal of Monetary Economics, 50*(1), 167–193.

Sachs, J. D., Ahluwalia, I. J., Amoako, K., Aninat, E., Cohen, D., Diabre, Z., et al. (2001). *Investing in Health for Economic Development*: Report of the Commission on Macroeconomics and Health.

Schultz, T. (1961). Investment in human capital. *American Economic Review, 51*(1), 1–17.

Schultz, T. (2005). Productive benefits of health: Evidence from low-income countries.

Schultz, T. P. (2003). Human capital, schooling and health. [doi: DOI: 10.1016/S1570-677X(03)00035-2]. *Economics & Human Biology, 1*(2), 207–221.

Schultz, T. P., & Tansel, A. (1997). Wage and labor supply effects of illness in Cote d'Ivoire and Ghana: Instrumental variable estimates for days disabled. *Journal of development economics, 53*(2), 251–286.

Schultz, T. P., & Tansel, A. (1997). Wage and labor supply effects of illness in Cote d'Ivoire and Ghana: Instrumental variable estimates for days disabled. *Journal of development economics, 53*(2), 251–286.

Schultz, T. W. (1961). Investment in human capital. *American Economic Review, 51*(1), 1–17.

Schultz, T. W. (1962). Reflections on investment in man. *Journal of political economy, 70*(5), 1–8.

Schultze, C. L. (1983). Industrial policy: A dissent. *The Brookings Review,* *2*(1), 3–12.

Sen, A. (1999). *Development as freedom*: Oxford University Press.

Sequeira, T. N. (2008). On the effects of human capital and R&D policies in an endogenous growth model. [doi: DOI: 10.1016/j.econmod.2008.01.002]. *Economic Modelling, 25*(5), 968–982.

Serageldin, I., & Grootaert, C. (1999). Defining social capital: an integrating view. In P. Dasgupta & I. Serageldin (Eds.), *Social Capital: A Multfaceted Perspective.* Washington D.C: Worldbank.

Sixsmith, J., Boneham, M., & Goldring, J. (2001). *The relationship between social capital, health and gender: A case study of Socially deprived Community.* London: Health Development Agency.

Smith, V. L. (1975). The primitive hunter culture, Pleistocene extinction, and the rise of agriculture. *Journal of Political Economy,* 727–755.

Sobel, J. (2002). Can we trust Social Capital. *Journal of Economic Literature, 40*(1), 139–154.

Solow, R. M. (1999). Notes on social capital and economic performance. In P. Dasgupta & I. Serageldin (Eds.), *Social Capital: A Multfaceted Perspective.* Washington D.C: Worldbank.

Sorrenson, M. (1967). *Land reform in the Kikuyu country: a study in government policy*: Nairobi. Oxford, U. P.

Spagat, M. (2006). Human capital and the future of transition economies. [doi: DOI: 10.1016/j.jce.2005.11.002]. *Journal of Comparative Economics, 34*(1), 44–56.

Stiglitz, J. E. (1999). Formal and informal institutions. In P. Dasgupta & I. Serageldin (Eds.), *Social Capital: A Multfaceted Perspective* (pp. 59–68). Washington D.C: Worldbank.

Strauss, J., & Thomas, D. (1995). Human resources: Empirical modeling of household and family decisions. *Handbook of development economics, 3*, 1883–2023.

Strauss, J., & Thomas, D. (1998). Health, Nutrition, and Economic Development. *Journal of Economic Literature, 36*(2).

Stroup, R. H., & Hargrove, M. B. (1969). Earnings and education in rural South Vietnam. *Journal of Human Resources*, 215–225.

Syverson, C. (2011). What Determines Productivity. *Journal of Economic Literature, 49*(2), 326–365.

Tamura, R. (2006). Human capital and economic development. [doi: DOI: 10.1016/j.jdeveco.2004.12.003]. *Journal of Development Economics, 79*(1), 26–72.

Temple, J. (1999). A positive effect of human capital on growth. *Economics Letters, 65*(1), 131–134.

Thirwall, A. P. (2006). *Growth and Development: With Special Reference to Developing Economies* (8 ed.). New York: Palgrave McMillan.

Thomas, D., & Strauss, J. (1997). Health and wages: Evidence on men and women in urban Brazil. *Journal of Econometrics, 77*(1), 159–185.

Timmer, C. (2002). Agriculture and economic development. In B. L. Gardner & G. C. Rausser (Eds.), *Handbook of agricultural economics* (Vol. 2, pp. 1487–1546): Elsevier.

Todaro, M. P., & Smith, S. C. (2009). *Economic Development* (10 ed.). Essex, England: Addison-Wesley.

Tran, D. T. (2009). *Education and Growth: A study of Vietnam's provinces*. Paper presented at the International Conference on Knowledge-based economy and global management.

Tran, D. T., & Do, N. T. (2007). The role of human capital in economic growth: A study of Vietnam's provinces in the period 2000–2004. From http://www.gsneu.edu.vn.

Tran, T. B., Grafton, R. Q., & Kompas, T. (2009). Institutions matter: The case of Vietnam. *Journal of Socio-Economics, 38*(1), 1–12.

Tran, T. Q. (2009). Sudden Surge in FDI and Infrastructure Bottlenecks: The Case in Vietnam. *ASEAN Economic Bulletin, 26*(1), 58–76.

Tuong, P. V. (2007). *Harmful habits and customs to health in Tay Nguyen ethnics*: Ministry of Health, Hanoi School of Public Health.

UNDP. (2009). *Human Development Report: Overcoming Barriers: Human Mobility and Development*: Palgrave Macmillan.

Uphoff, N. (1999). Understanding social capital: learning from the analysis and experience of participation. In P. Dasgupta & I. Serageldin (Eds.), *Social Capital: A Multfaceted Perspective* (pp. 215–249). Washington D.C.: Worldbank.

Valadkhani, A. (2005). Modelling Demand for Broad Money in Australia. *Australian Economic Papers, 44*(1), 47–64.

Van Arkadie, B., & Mallon, R. (2003). *Vietnam: a transition tiger?* Canberra, Australia: Asia Pacific Press.

Van de Walle, D., & Cratty, D. (2004). Is the emerging non farm market economy the route out of poverty in Vietnam? *Economics of Transition, 12*(2), 237–274.

Van der Gaag, M. (2005). *Measurement of individual social capital*. University of Groningen.

Van Trinh, L. T., Gibson, J., & Oxley, L. (2005). Measuring the stock of human capital in New Zealand. [doi: DOI: 10.1016/j.matcom.2005.02.007]. *Mathematics and Computers in Simulation, 68*(5–6), 484–497.

Vo, T. T., & Nguyen, A. D. (2009). Vietnam after two years of WTO accession: What lessons can be learnt? *ASEAN Economic Bulletin, 26*(1), 115–135.

Walle, D. (2003). Are returns to investment lower for the poor? Human and physical capital interactions in rural Vietnam. *Review of Development Economics, 7*(4), 636–653.

Ware, J., Davies-Avery, A., & Donald, C. A. (1978). Conceptualization and measurement of health for adults in the health insurance study: Vol. V, general health perceptions. *Santa Monica, CA: Rand Corporation, 28.*

Wei, H. (2004). Measuring the stock of human capital for Australia. *Australian Bureau of Statistics Working Paper.*

Weisbrod, B. A. (1961). The valuation of human capital. *Journal of Political Economy, 69*(5), 425–436.

Weisbrod, B. A. (1962). Education and investment in human capital. *Journal of political economy, 70*(5), 106–123.

Weisdorf, J. L. (2005). From foraging to farming: Explaining the Neolithic revolution. *Journal of Economic Surveys, 19*(4), 561–586.

Weitz, R. (1971). *From peasant to farmer. A revolutionary strategy for development.*

Wickens, C. (1924). *Human capital.*

Wiegersma, N. (1988). *Vietnam: peasant land, peasant revolution: patriarchy and collectivity in the rural economy*: St. Martin's Press.

Wilkinson, R. (1996). *Unhealthy Societies: the afflictions of inequality.* London: Routledge.

Willis, R. J., & Rosen, S. (1979). Education and self-selection. *Journal of Economic Perspectives, 87*(2), 7–36.

Wolff, E. (2000). Human capital investment and economic growth: exploring the cross-country evidence. *Structural Change and Economic Dynamics, 11*(4), 433–472.

Woolcock, M. (1999). *Social Capital: The state of the Notions.* Paper presented at the Social Capital: Global and Local Perspective.

Woolcock, M. (2001). The place of social capital in Understanding Social and Economic Outcomes. *ISUMA Canadian Journal of Policy Research, 2*(1), 11–17.

WorldBank. (1993). *The East Asian Miracle: Economic Growth and Public Policy*: Oxford University Press New York.

WorldBank. (1999). *World development report 2000: attacking poverty* (No. 0305–750X): Oxford, Eng.; New York: Pergamon Press, 1973.

WorldBank. (2001). *World development report 2002: building institutions for markets*. New York: The World Bank.

WorldBank. (2007). *Vietnam Development Report 2008: Social Protection*. Ha Noi: The World Bank.

WorldBank. (2008). *World Development Report: Agriculture for Development*: The World Bank.

Wößmann, L. (2003). Specifying human capital. *Journal of Economic Surveys, 17*(3), 239–270.

Yamamura, E. (2009). Dynamics of social trust and human capital in the learning process: The case of the Japan garment cluster in the period 1968–2005. [doi: DOI: 10.1016/j.jebo.2009.06.001]. *Journal of Economic Behavior & Organization, 72*(1), 377–389.

Yew, S. L., & Zhang, J. (2009). Optimal social security in a dynastic model with human capital externalities, fertility and endogenous growth. [doi: DOI: 10.1016/j.jpubeco.2008.10.005]. *Journal of Public Economics, 93*(3–4), 605–619.

Zepeda, L. (2001). Agricultural investment, production capacity and productivity. In L. Zepeda (Ed.), *Agricultural investment and productivity in developing countries*: Food & Agriculture Organization of the UN (FAO).

APPENDIX

Appendix A. Statistic Summarize

Table 5.2(a) Summarize statistic of variables in VHLSS 2002

Variables	Observations	Mean	Standard of Deviation	Min	Max	Descriptions
ln_income02	10278	8.74	1.10	2.91	12.48	Natural log of agricultural income in year 2002
ln_land02	10041	8.23	1.11	2.30	12.46	Natural log of agricultural land area in year 2002
ln_food02	10278	8.84	0.48	6.57	10.94	Natural log of agricultural households expenditure on food in year 2002
ln_health02	10278	6.55	1.16	0.22	11.16	Natural log of agricultural households expenditure on health care in year 2002
ln_edu02	7316	6.42	1.10	0.22	10.43	Natural log of agricultural household expenditure on education in year 2002
schooling	10278	5.45	2.74	0	12	Years of schooling of household head and members reported in year 2002
exp02	10278	26.57	11.76	4.5	88	Experience = Ages – Schooling – 5
exp022	10278	844.75	858.56	20.25	7744	Experience Square
life02	10278	68.59	3.96	55.4	79.1	Life expectancy at birth

Table 5.2(b) Summarize statistic of variables in VHLSS 2004

Variables	Observations	Mean	Standard of Deviation	Min	Max	Descriptions
ln_income04	3386	9.84	0.64	7.35	12.71	Natural log of agricultural income in year 2004
ln_land04	3386	7.89	0.93	4.78	11.51	Natural log of agricultural land area in year 2004
ln_food04	3383	6.59	0.63	1.92	8.97	Natural log of agricultural households expenditure on food in year 2004
ln_health04	3386	6.39	1.26	1.53	11.84	Natural log of agricultural households expenditure on health care in year 2004
ln_edu04	2503	6.51	1.12	1.56	9.79	Natural log of agricultural household expenditure on education in year 2004
schooling04	3386	6.72	2.77	0	12	Years of schooling of household head and members reported in year 2004
exp	3386	27.00	11.38	6.25	83	Experience = Ages – Schooling – 5
exp2	3386	859.18	803.86	39.06	6889	Experience Square
life04	3386	68.94	3.91	56.4	79	Life expectancy at birth

INVESTMENT AND AGRICULTURAL DEVELOPMENT IN DEVELOPING COUNTRIES 371

Figure 5.2. Frequency distributions for log agricultural households and food expenditure income in 2002

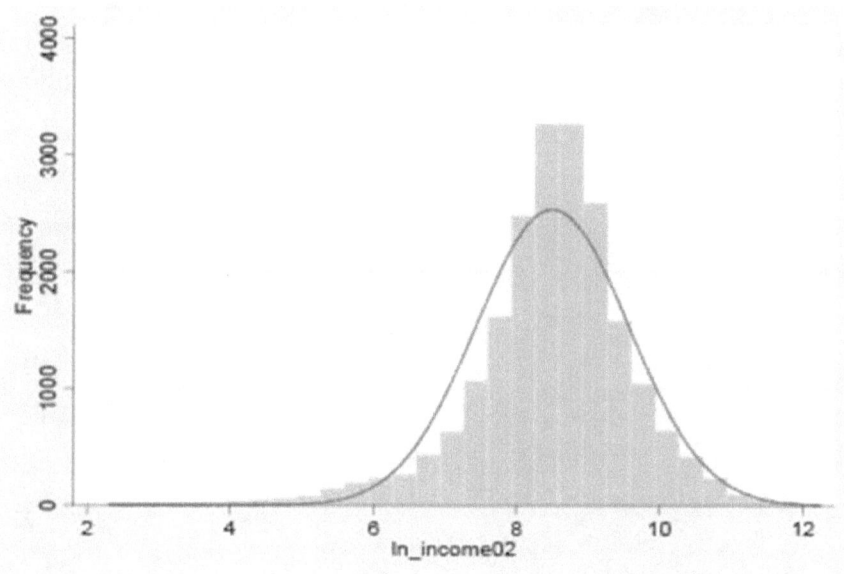

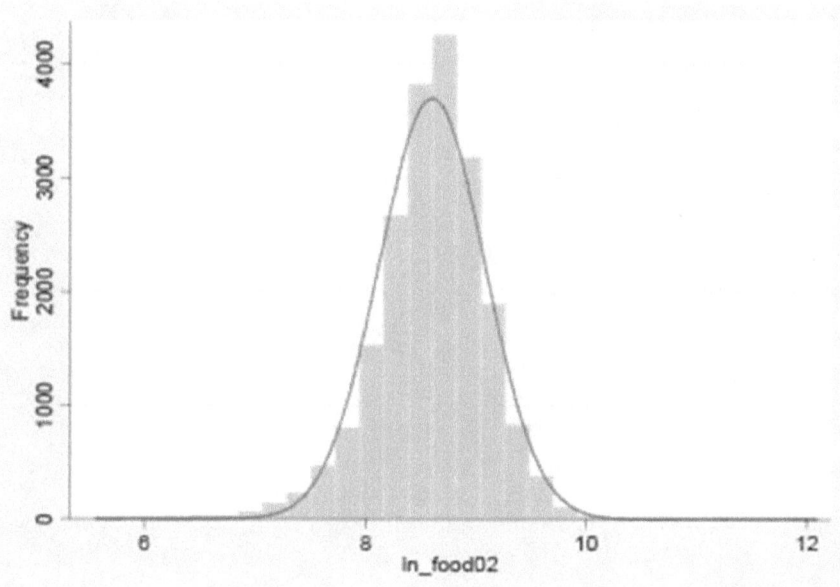

Figure 5.2. Frequency distributions for log agricultural households and food expenditure income in 2004

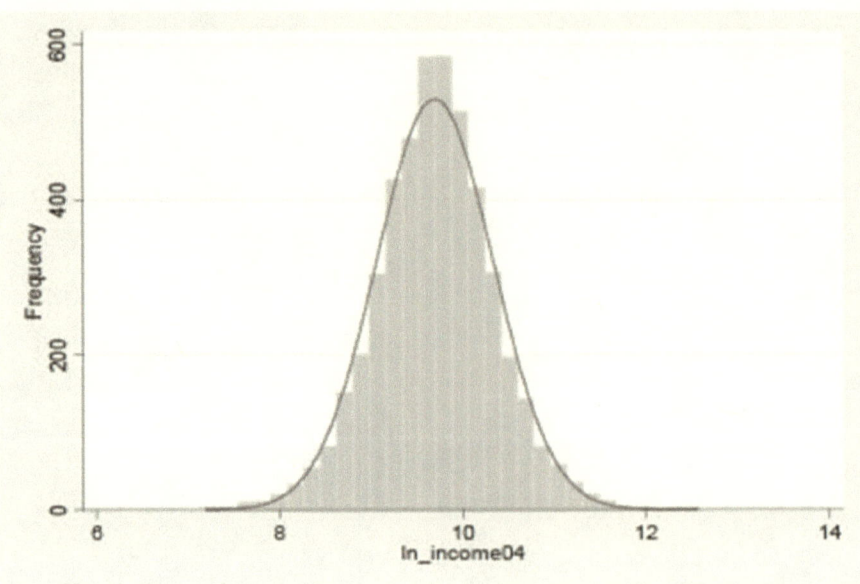

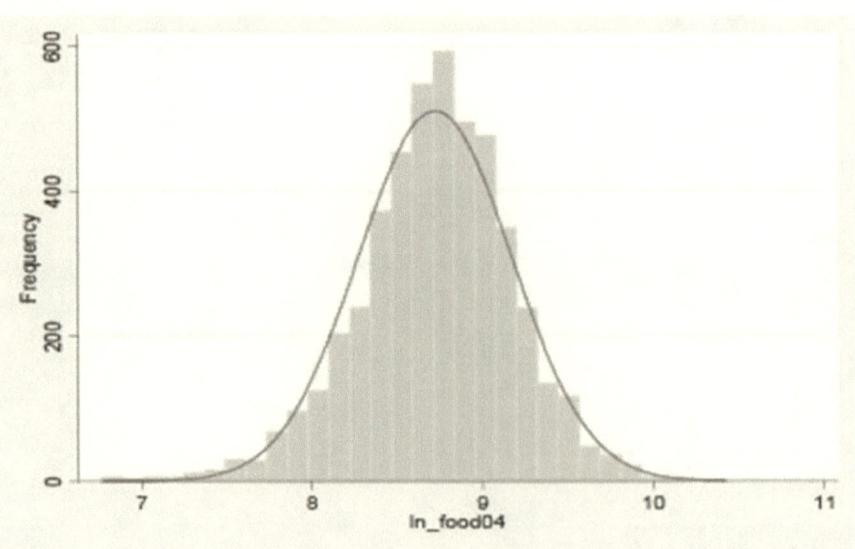

Appendix B. Empirical Results

Table 5.3. Estimation result by using VHLSS 2002

Variables	Methods of regression			
	OLS	GLS/RE	IV-2SLS	G2SLS – IV/RE
Log of land area	0.74*** (0.0074)	0.74*** (0.0072)	0.73*** (0.0071)	0.73*** (0.0071)
Log of food expenditure	0.19*** (0.020)	0.24*** (0.020)	0.25*** (0.0177)	0.24*** (0.0177)
Log of health expenditure	0.015*** (0.006)	0.0169*** (0.0059)	0.0053 (0.0062)	0.0133** (0.00601)
Log of education expenditure	0.0507*** (0.0079)	0.0516*** (0.0077)		
Years of Schooling	0.013*** (0.0032)	0.00975*** (0.0031)	0.0199*** (0.0028)	0.0190*** (0.00277)
Experience	−0.0011 (0.0038)	0.00091 (0.0037)	−0.00202 (0.00392)	−0.00021 (0.00376)
Experience square	Nearly 0 and not significant	−.0000181 (.0000621)	.0000247 (.0000665)	Nearly 0 and not significant
Life expectancy	0.022*** (0.0032)	0.017 (0.0112)	0.064*** (0.00299)	0.060*** (0.00467)
Household size	−0.0108** (0.0054)	−0.0157*** (0.0053)		
Gender (female=1)	−0.1723*** (0.0269)	−0.1422** (0.0676)	−0.4169*** (0.02651)	−0.3845*** (0.0332)
Red River Delta	0.054** (0.0231)	0.134 (0.0908)		
North East	−0.0899*** (0.0272)	−0.0585 (0.1028)		
North West	−0.3269*** (.0370)	−0.2677* (.1495)		
North Central Coast	−.3186*** (0.0268)	−.2873*** (0.1104)		
South Central Coast	−0.2504*** (.0298)	−0.1577 (.1100)		
Central Highlands	−0.3178*** (0.0338)	−0.3115** (0.1458)		
South East	−0.2497*** (0.0295)	−0.1918* (0.0996)		
Constant	−0.7164** (0.2884)	−1.032 (0.8012)	−3.931*** (0.2585)	−3.703*** (0.3551)
Observations	7172	7172	7172	7172

	Methods of regression			
Variables	OLS	GLS/RE	IV-2SLS	G2SLS – IV/RE
R²	0.6713	within = 0.6419 between = 0.7360 overall = 0.6701	0.6496	within = 0.6393 between = 0.6382 overall = 0.6513
R² adjusted	0.6705		0.6492	
F	859.47		1676.28	
Wald χ²		12890.54	12201.60	12855.94
sigma_u		0.2076		0.2036
sigma_e		0.5412		1.5858
Rho		0.1283		0.01621
Note: - numbers in parentheses are standard of errors. - *, ** and *** are statistic significant at 10%, 5%, and 1% respectively.				

Source: Author's calculation

Table 5.4. Estimation result by using VHLSS 2004

	Methods of regression			
Variables	OLS	GLS/RE	IV-2SLS	G2SLS – IV/RE
Log of land area	0.0821*** (0.0123)	0.0815*** (0.0124)	0.1915 *** (0.0106)	0.1411*** (0.0119)
Log of food expenditure	0.4372*** (0.0164)	0.4468*** (0.0166)	0.4806*** (0.0160)	0.5019*** (0.0165)
Log of health expenditure	0.0583*** (0.0073)	0.0521*** (0.0071)	0.0857*** (0.00777)	0.0702*** (0.00745)
Log of education expenditure	0.0804*** (0.0094)	0.0789*** (0.0092)		
Years of schooling	0.0375*** (0.00487)	0.0395*** (0.00489)	0.0232*** (0.00456)	0.0375*** (0.00456)
Experience	0.0032 (0.0048)	0.0039 (0.00467)	0.0046 (0.00519)	0.00536 (0.0049)
Experience square	−0.0000108 (0.0000795)	−0.0000175 (0.0000774)	−0.0000718 (0.0000861)	−0.0000642 (0.0000812)
Life expectancy	0.0026 (0.0043)	0.00611 (0.00903)	0.0339*** (0.0041)	0.0370*** (0.00791)
Household size	0.0557*** (0.0060)	0.0534*** (0.00588)		
Gender (female=1)	−0.0477 (0.0356)	−0.0694 (0.0576)	−0.2387*** (0.0376)	−0.2691*** (0.0532)
Red River Delta	−0.2992*** (0.0352)	−0.3199*** (0.0714)		
North East	−0.4689*** (0.0415)	−0.4652*** (0.0807)		
North West	−0.3931*** (0.0474)	−0.3789*** (0.1055)		

Variables	Methods of regression			
	OLS	GLS/RE	IV-2SLS	G2SLS – IV/RE
North Central Coast	−0.4845*** (0.0377)	−0.4775*** (0.0843)		
South Central Coast	−0.2736*** (0.0386)	−0.2508*** (0.0843)		
Central Highlands	−0.1277** (0.0504)	−0.1389 (0.1062)		
South East	−0.0923*** (0.0447)	−0.0455 (0.0814)		
Constant	4.957*** (0.3447)	4.704*** (0.6487)	2.132*** (0.3120)	2.185*** (0.5522)
Observations	2502	2502	2502	2502
R²	0.5259	within = 0.4715 between = 0.7548 overall = 0.5248	0.4418	within = 0.4302 between = 0.5060 overall = 0.4275
R² adjusted	0.5226		0.4400	
F	162.07		252.15	
Wald χ²		2323.89		1909.63
sigma_u		0.1483		0.1395
sigma_e		0.4038		0.4411
Rho		0.1189		0.0910

Note: - numbers in parentheses are standard of errors.
- *, **, and *** are statistic significant at 10%, 5%, and 1%, respectively.

Table 5.5 Estimate the speed of knowledge diffusion by using VHLSS 2002, 2004

Variables	Methods of regression			
	IV-2SLS			
	(1)	(2)	(3)	(4)
Growth of land area	0.2515*** (0.0246)	0.2604*** (0.0248)	0.2610*** (0.0248)	0.2611*** (0.0249)
Growth of food expenditure	0.2317*** (0.0301)	0.1993*** (0.0307)	0.2039*** (0.0308)	0.2067*** (0.0309)
Growth of health expenditure	0.0211* (0.0112)	0.0208* (0.0113)	0.0206* (0.0113)	0.0204* (0.0113)
Schooling		0.0355*** (0.0069)	0.0391*** (0.0077)	0.0391*** (0.0078)
Experience			0.0021 (0.00166)	0.00515 (0.0069)
Experience square				−0.000046 (0.000099)
Life expectancy	0.0417 (0.0918)	0.1662*** (0.0952)	0.1703* (0.0957)	0.1638* (0.0959)
Speed of knowledge diffusion	0.1566*** (0.0084)	0.1183 *** (0.0112)	0.1112* (0.0132)	0.1079*** (0.0165)

Observations	1793	1793	1793	1793
F	778.05	622.95	518.07	443.34
R^2	0.1654	0.1600	0.1574	0.1560
Adj R^2	0.1635	0.1576	0.1546	0.1527
Tests of endogeneity H_0: variables are exogenous				
Durbin (score) $\chi^2(1)$	364.103	408.212	390.719	370.578
Wu-Hausman $F(1,1788)$	455.608	526.777	497.635	465.039
Note: - numbers in parentheses are standard of errors. - *, **, and *** are statistic significant at 10%, 5%, and 1%, respectively.				

Table 5.6. The effect of investing in health on income of agricultural households in 2002, by regions

Variables	Red River Delta	North East	North West	North Central Coast	South Central Coast	Central Highlands	South East	Mekong River Delta	All regions
Log of land area	0.8588*** (0.0125)	0.6302*** (0.0153)	0.5205*** (0.0198)	0.7151*** (0.0168)	0.6260*** (0.0239)	0.6469*** (0.0227)	0.6252*** (0.0296)	0.9343*** (0.0144)	0.7015*** (0.00619)
Log of food expenditure	0.1191*** (0.0247)	0.3646*** (0.0391)	0.2780*** (0.0582)	0.1725*** (0.0426)	0.2373*** (0.0582)	0.4508*** (0.0594)	0.3430*** (0.0907)	0.2665*** (0.0383)	0.2925*** (0.01666)
Log of health expenditure	0.00007 (0.00764)	0.0055 (0.0107)	0.0613*** (0.0193)	0.0433*** (0.0139)	-0.00611 (0.02023)	0.1162*** (0.0202)	-0.0125 (0.0292)	-0.01235 (0.01361)	0.0322*** (0.00544)
Years of schooling	0.01313*** (0.0036)	0.00957*** (0.00505)	0.0318*** (0.00886)	0.0131** (0.0061)	0.0195** (0.00914)	0.0502*** (0.00912)	0.0277* (0.0142)	-0.0032 (0.00677)	0.0297*** (0.00247)
Experience	0.0080** (0.0029)	-0.000988 (0.00527)	0.0184 (0.01146)	0.00706 (0.0051)	0.0189*** (0.0067)	-0.0027 (0.00860)	0.0155 (0.0137)	-0.0052 (0.00569)	0.00319 (0.00223)
Experience square	-0.00013*** (0.000040)	-0.0000109 (0.0000802)	-0.000255 (0.000183)	-0.00013 (0.00007)	-0.00022** (0.000086)	0.0000999 (0.000122)	-0.000272 (0.000199)	0.0000149 (0.0000797)	-0.0000491 (0.0000312)
Life expectancy	-0.01460** (0.00652)	0.0236*** (0.00451)	-0.00387 (0.01676)	0.0489*** (0.0089)	-0.0291** (0.0114)	0.00194 (0.00175)	-0.03350 (0.0198)	0.0514*** (0.0121)	0.0145*** (0.001119)
Gender (female=1)	0.0709 (0.0442)	-0.1468*** (0.0456)	0.00044 (0.11816)	-0.38103*** (0.0669)	0.1167 (0.0826)	-0.01033 (0.0652)	0.1754 (0.14288)	-0.38756*** (0.0751)	-0.0878*** (0.01713)
Household size	-0.0044 (0.00876)	-0.0111 (0.0105)	-0.00720 (0.0135)	0.0205 (0.0142)	0.0175 (0.0186)	-0.0411** (0.0172)	0.0683** (0.0269)	-0.0363*** (0.01096)	-0.0128** (0.00512)
Constant	1.865*** (.5069)	-1.219*** (.4649)	1.329 (1.241)	-2.558*** (.7301)	2.963*** (.8717)	-1.694*** (.540)	2.3966* (1.415)	-4.3886*** (0.9389)	-.9101*** (0.1618)
Observations	2252	1604	491	1237	802	884	759	2012	10041
R-squared	0.7450	0.6050	0.6914	0.6755	0.5824	0.5753	0.4588	0.7121	0.6431
Adj. R-squared	0.7440	0.6028	0.6856	0.6731	0.5777	0.5709	0.4522	0.7108	0.6428
F	727.70	271.30	119.75	283.76	122.73	131.54	70.54	550.21	2008.10

Note: - number in parentheses is standard of error.
- *, **, and *** are statistic significant at 10%, 5%, and 1%, respectively.

Source: Author's calculation

Table 5.7. The effect of investing in health on income of agricultural households in 2004, by regions

Variables	Red River Delta	North East	North West	North Central Coast	South Central Coast	Central Highlands	South East	Mekong River Delta	All regions
Log of land area	0.0102 (0.0242)	0.0133 (0.0298)	0.1664*** (0.0344)	0.0675** (0.0323)	0.1182*** (0.0318)	0.0666 (0.0431)	0.2048*** (0.0438)	0.2087*** (0.0244)	0.1864*** (0.0091)
Log of food expenditure	0.5949*** (0.0265)	0.520*** (0.0415)	0.3521*** (0.0505)	0.5319*** (0.0392)	0.3726*** (0.0422)	0.3153*** (0.0452)	0.2938*** (0.0581)	0.3214*** (0.0352)	0.4337*** (0.0136)
Log of health expenditure	0.0329*** (0.0120)	0.0778*** (0.0144)	0.1123*** (0.0190)	0.0763*** (0.0163)	0.0465** (0.0197)	0.0678** (0.0269)	0.1652*** (0.0281)	0.0501*** (0.0178)	0.0858*** (0.0064)
Years of schooling	0.0390*** (0.0390)	0.0593*** (0.0086)	0.0629*** (0.0113)	0.0462*** (0.0100)	0.0756*** (0.0117)	0.0489*** (0.0134)	0.0413** (0.0179)	0.0598*** (0.0106)	0.0404*** (0.0037)
Experience	0.0083* (0.0048)	0.0065 (0.0082)	0.0055 (0.0116)	0.0107 (0.0087)	0.01743* (0.00906)	0.0047 (0.0132)	-0.00477 (0.0157)	0.0184* (0.0095)	0.0111*** (0.00309)
Experience square	-0.00017 (0.000066)	-0.000063 (0.00013)	0.000061 (0.000175)	-0.000161 (0.00013)	-0.000217* (0.000117)	-0.000196 (0.000185)	0.000102 (0.00023)	-0.00027* (0.00014)	-0.000187*** (0.000044)
Life expectancy	-0.0113 (0.0115)	0.0191*** (0.00653)	0.00603 (0.0188)	-0.00998 (0.0121)	0.0184 (0.0132)	-0.00284 (0.00936)	0.0722*** (0.02198)	-0.0359** (0.0166)	0.02402*** (0.00273)
Gender (female =1)	0.0828 (0.0743)	-0.1522** (0.07086)	0.00775 (0.1574)	0.0510 (0.0942)	-0.1053 (0.0893)	-0.1014 (0.1124)	-0.5165*** (0.1455)	0.1822* (0.1042)	-0.1371*** (0.0271)
Household size	0.0972*** (0.0121)	0.0589*** (0.0135)	0.0456*** (0.0119)	0.0772*** (0.0158)	0.0896*** (0.0154)	0.0603*** (0.0159)	0.0589*** (0.0181)	0.0490*** (0.0127)	0.05900*** (0.00525)
Constant	5.629*** (0.8652)	3.5213 (0.5610)	4.193*** (1.366)	4.997*** (0.927)	3.706*** (0.9428)	6.522*** (0.7236)	-.1375 (1.5493)	7.538*** (1.166)	2.641*** (.2104)
Observations	929	565	235	443	320	198	144	549	3383
R-squared	0.5900	0.4392	0.5768	0.5369	0.6271	0.6007	0.6533	0.5020	0.5312
Adj. R-squared	0.5860	0.4301	0.5599	0.5272	0.6162	0.5816	0.6300	0.4937	0.5300
F	146.96	48.30	34.08	55.77	57.91	31.43	28.06	60.38	424.71

Note: - number in parentheses is standard of error.
- *, **, and *** are statistic significant at 10%, 5%, and 1%, respectively.

Source: Author's calculation

Appendix C

Table 5.8. Estimates of the effect of health on economic growth [a]

Study	Health measure (in logs)	Coefficient (standard error)	Growth effect of increasing life expectancy by 5 years	Data	Estimator	Other covariates (all papers have the log of initial income per capita or per worker)
Barro (1996)	Life expectancy	0.042 (0.014)	0.33	Three periods 1965–1975, n=80; 1975–1985, n=87; 1985–1990, n=84	3SLS using lagged values of some regressions as instruments, period random effects	Male secondary and higher schooling, log(GDP) × male schooling, log fertility rate, government consumption ratio, rule of law index, terms of trade change, democracy index, democracy index squared, inflation rate, continental dummies
Barro and Lee (1994)	Life expectancy	0.073 (0.013)	0.58	Two periods n=85 for 1965–1975, n=95 for 1975–1985	SUR with country random effects	Male and female secondary schooling, 1/GDP, G/GDP, log(1+ black market premium), revolution
Barro and Sala-I-Martin (1995)	Life expectancy	0.058 (0.013)	0.46	Two periods n=87 for 1965–1975, n=97 for 1975–1985	SUR with country random effects	Male and female secondary and higher education, log(GDP) × human capital, public spending on education/GDP, investment/GDP, government consumption/GDP, log(1+ black market premium), growth rate in terms of trade
Bhargava, Jamison, Lau, and Murray (2001)	Adult survival rate ASR × log(GDPC)	0.358 (0.114) −0.048 (0.016)	NA	25-year panel at 5-year intervals, 1965–90, n=92	Dynamic random effects	Tropics, openness, log fertility, log (investment/GDP)

Study	Health measure (in logs)	Coefficient (standard error)	Growth effect of increasing life expectancy by 5 years	Data	Estimator	Other covariates (all papers have the log of initial income per capita or per worker)
Bloom, Canning and Malaney (2000)	Life expectancy	0.063 (0.016)	0.50	25-year panel at 5-year intervals, 1965–1990, n=391	Pooled OLS	GDP per worker, tropics, landlocked, institutional quality, openness, log of years of secondary schooling, population growth, working age population growth, log ratio of working-age to total population, population density, period dummies
Bloom and Malaney (1998)	Life expectancy	0.027 (0.107)	0.21	25 year cross-section, 1965–1990, n=77	OLS	Population growth, growth of economically active population, log years of secondary schooling, natural resource abundance, openness, institutional quality, access to ports, average government savings, tropics, ratio of coastline distance to land area
Bloom et.al (1999)	Life expectancy	0.019 (0.012)	0.15	25-year cross-section, 1965–1990, n=80	2SLS	Log of ratio of total population to working-age population, tropics, log of years of secondary schooling, openness, institutional quality, population growth rate, working-age population growth rate
Bloom and Sachs (1998)	Life expectancy	0.037 (0.011)	0.29	25-year cross-section, 1965–1990, n=65	OLS	Log secondary schooling, openness, institutional quality, central government deficit, percentage area tropics, log coastal population density, log inland population density, total population growth rate, working age population growth rate, African dummy
Bloom and Williamson (1998)	Life expectancy	0.040 (0.010)	0.32	25 year cross-section, 1965–1990, n=78	OLS	Population growth rate, working-age population growth rate, log years of secondary schooling, natural resource abundance, openness, institutional quality, access to port, average government saving rates, tropic dummy, ratio of coastline to land area

Study	Health measure (in logs)	Coefficient (standard error)	Growth effect of increasing life expectancy by 5 years	Data	Estimator	Other covariates (all papers have the log of initial income per capita or per worker)
Caselli, Esquivel, and Lefort (1996)	Life expectancy	−0.001 (0.032)	0.00	25-year panel at 5-year intervals, 1965–1980, n=91	GMM (Arellano-Bond method)	Male and female schooling, I/GDP, G/GDP, black market premium, revolutions
Gallup and Sachs (2000)	Life expectancy	0.030 (0.009)	0.24	25 year cross section, 1965–1990, n=75	OLS	Years of secondary schooling, openness, quality of public institutions, population within 100 kilometers of the coast, malaria index 1966, change in malaria index from 1966 to 1994
Hamoudi and Sachs (1999)	Life expectancy	0.072 (0.020)	0.57	15-year cross-section, 1980–1995, n=78	OLS	Institutional quality, openness, net government savings, tropics land area, log coastal population density, population growth rate, working-age population growth rate, Africa dummy Openness, openness
Sachs and Warner (1997)	Life expectancy life expectancy squared	45.48 (18.49) −5.40 (2.24)	0.06	25 year cross-section, n=79	OLS	Openness, openness × log(GDP), land-locked, government saving, tropical climate, institutional quality, natural resource exports, growth in economically active population minus population growth
Bloom, Canning and Sevilla (2004)	Life expectancy	0.04 (0.019)	0.20	30-year panel at 10-year intervals, 1960–1990, n=147	2SLS	Log capital per worker, average years of schooling, average experience, average experience squared, %tropical area, governance

ASR: adult survival rate; GDP: gross domestic product; GMM: generalized method of moments; OLS: ordinary least squares; 2SLS two stages least square; 3SLS: three stages least square; SUR: seemingly unrelated regression.

[a] The growth effects of a five year increase in life expectancy are calculated for a country with a life expectancy of 63, the average life expectancy in developing countries in 1990.

Source: Bloom et.al (2004).

Appendix D

Table 7.11. Statistical summary of variables

Name of variables	2002					2004					2006				
	N	Mean	S.E	Min	Max	N	Mean	S.E	Min	Max	N	Mean	S.E	Min	Max
Log of working hour	28456	1.74	0.39	0	2.77	1047	1.84	0.24	0.69	2.48	1717	1.86	0.25	0.69	2.39
Log of contributed fund	28456	1.27	1.85	0	7.60	1047	4.28	1.08	0.69	8.07	1717	4.56	1.09	0.92	7.82
Log of pocket money for child	28456	0.68	1.74	0	8.18	1047	4.79	1.26	1.09	8.37	1717	5.19	1.25	1.38	8.18
Log of holiday outdoor food expenditure	28456	4.08	0.90	1.09	7.82	1047	5.17	0.84	2.30	7.71	1717	6.52	0.63	3.58	8.47

www.ingramcontent.com/pod-product-compliance
Lightning Source LLC
Chambersburg PA
CBHW020722180526
45163CB00001B/73